THE SOUND OF EXCLUSION

Latinx Pop Culture

SERIES EDITORS

Frederick Luis Aldama and Arturo J. Aldama

THE SOUND OF EXCLUSION

NPR and the Latinx Public

Christopher Chávez

THE UNIVERSITY OF
ARIZONA PRESS
TUCSON

The University of Arizona Press
www.uapress.arizona.edu

We respectfully acknowledge the University of Arizona is on the land and territories of Indigenous peoples. Today, Arizona is home to twenty-two federally recognized tribes, with Tucson being home to the O'odham and the Yaqui. Committed to diversity and inclusion, the University strives to build sustainable relationships with sovereign Native Nations and Indigenous communities through education offerings, partnerships, and community service.

ISBN-13: 978-0-8165-4276-5 (paperback)

Cover design by Leigh McDonald
Cover photo by Marc Schulte/Unsplash
Typeset by Sara Thaxton in 10.5/13.5 Warnock Pro with Futura Std

Publication of this book is made possible in part by support from the Oregon Humanities Center at the University of Oregon, and by the proceeds of a permanent endowment created with the assistance of a Challenge Grant from the National Endowment for the Humanities, a federal agency.

Library of Congress Cataloging-in-Publication Data
Names: Chávez, Christopher, author.
Title: The sound of exclusion : NPR and the Latinx public / Christopher Chávez.
Other titles: Latinx pop culture.
Description: Tucson : University of Arizona Press, 2021. | Series: Latinx pop culture | Includes bibliographical references and index.
Identifiers: LCCN 2021021355 | ISBN 9780816542765 (paperback)
Subjects: LCSH: National Public Radio (U.S.) | Hispanic Americans and mass media. | Hispanic Americans—Press coverage.
Classification: LCC HE8697.95.U6 C53 2021 | DDC 384.54089/68073—dc23
LC record available at https://lccn.loc.gov/2021021355

Printed in the United States of America
♾ This paper meets the requirements of ANSI/NISO Z39.48-1992 (Permanence of Paper).

For Judy, Alexandra, and Daniela. And for my mother.

CONTENTS

ACKNOWLEDGMENTS

I began working on this book several months before Donald Trump was elected to office and completed the manuscript, four years later, during the final days of his administration. While this book serves as a historical perspective of how NPR has evolved over time, it becomes clear that the book is also a reflection of how Latinx journalists were navigating their profession at this particular moment in time, in which anti-Latinx rhetoric was coming from the nation's most powerful.

An evaluation of public radio would not have been possible without the help of a number of industry professionals and academics. I want to begin by thanking my colleagues in journalism at the University of Oregon, who shared their insights and connections, particularly Charlie Butler, Damian Radcliffe, and Dr. Regina Lawrence. But I am especially grateful to Dr. Peter Laufer, who has been a champion of this book from its inception. I have appreciated your guidance on writing and publishing and your insights on the field of journalism.

I am also grateful to my colleagues Dr. Gabriela Martínez, Dr. Kim Sheehan, Dr. Leslie Steeves, and Dr. Janet Wasko, who have been mentors during my time in Eugene. And to my colleagues in Latinx media studies across the academy—including Dr. Anghy Valdivia, Dr. Arlene Dávila, Dr. Mari Castañeda, Dr. Delores Ines-Casillas, Dr. Jillian Báez, Dr. Diana Leon-Boys, Dr. Claudia Bucciferro, Dr. Michael Lechuga, Dr. Litzy Galarza, Dr. Roberto Avant-Mier, Dr. Hector Amaya, and countless others who are building this field of study—you inspire me every day by the work you do.

I would also like to express gratitude to Dr. Frederick Luis Aldama, Dr. Arturo Aldama, and Kristen Buckles, for their belief in and investment in this project. And to Carla Alvarez, from the University of Texas Libraries, for her tremendous help in locating some of the book's historic images. Finally, I would like to thank the many public radio practitioners who participated in this study. Your keen insights and candid assessments of public radio's norms and practices were essential for understanding how NPR, and public media in general, can better serve the public interest.

THE SOUND OF EXCLUSION

INTRODUCTION

This erasure of Latinos by the national media is nothing new. For years, the marquee Sunday political talk shows have rarely featured Latinos. There is only one Latino on the *New York Times'* editorial board, and there is none on the *Washington Post'*s. NPR, where I work, recently had a period of time with no Latino reporters on its politics team before it made two hires.

— LOURDES GARCIA-NAVARRO, HOST OF NPR'S *WEEKEND EDITION SUNDAY**

On March 25, 2006, an estimated five hundred thousand demonstrators gathered in downtown Los Angeles to protest the passage of the Border Protection, Antiterrorism, and Illegal Immigration Control Act (HR 4437), proposed legislation that was designed to criminalize undocumented immigrants and those who give them aid. Known simply as the "Sensenbrenner Bill," the legislation prompted swift and immediate reaction from community organizations, trade unions, religious institutions, and residents alike.

Similar marches were reported in Denver, Phoenix, Atlanta, and other cities across the United States, but the Los Angeles march appeared especially strong. Said to be one the largest in the city's history, the protesters covered twenty-six blocks, marching along Spring Street, down Broadway and Main, finally making their way to City Hall, where Mayor Antonio Villaraigosa, the city's first Latinx mayor since 1872, addressed the crowd in Spanish (Watanabe and Becerra 2006a).

Looking at photographs of the event, it is hard not to notice the many ways in which Latinx protestors engaged in what Eco (1997) terms "semiological guerilla warfare," appropriating words, images, and symbols to openly challenge the state position while asserting their fundamental right to belong in a country that does not always wish to claim them. Some of the protestors wore T-shirts that read "We are America." Others waved American flags, gestures that were meant to be read less as overt displays of nationalism, but more as attempts to challenge popular conceptions of what it means to be American.

---

*From *The Atlantic*, August 9, 2019.

Others framed their protest in clear moral terms, carrying signs that read "We are imigrantes, not criminals" and "Ningún ser humano es ilegal!!" There were photographs of protestors wearing white, and others of clergy wearing handcuffs. In a show of solidarity, Los Angeles Catholic Archbishop Cardinal Mahony was photographed marching side by side with several Latinx youths. Mahony had set the stage two days earlier, writing an op-ed in the *New York Times* making the moral case against the bill. "While we gladly accept their taxes and sweat, we do not acknowledge or uphold their basic labor rights," Mahony wrote in the piece. "At the same time, we scapegoat them for our social ills and label them as security threats and criminals to justify the passage of anti-immigration bills" (Mahony 2006).

The 2006 marches have been the subject of scholarly interest, primarily with a focus on political mobilization (Beltrán 2010; Zepeda-Millán 2017). However, I want to draw attention to the journalistic coverage of the event, and in particular the disparities in coverage between different kinds of news organizations. Overall, journalists across the country appeared to be taken aback by the sheer size and diversity of the crowd in downtown Los Angeles. This is surprising, given that in 2006 Los Angeles County was home to the largest Latinx community in the nation, numbering over 4.7 million (U.S. Census 2006). But there was also confusion over how to cover the event. To get a handle on the story, national news media outlets dispatched reporters to interview protesters. Translators were employed en masse to make Spanish-language testimonies comprehendible to an intended audience of English monolingual listeners.

For its part, National Public Radio (NPR), the country's largest publicly funded radio network, appeared to approach the protests as if it were a form of war correspondence. That evening on *All Things Considered*, NPR's flagship news program, host Debbie Elliott attempted to provide context for the marches, beginning her segment by playing an audio excerpt from President George W. Bush's Saturday Radio Hour, in which he framed the immigration debate as a need to balance national security with serving the economic needs of businesses. To integrate the Latinx perspective, however, Elliott curiously turned to correspondent Carrie Kahn who described her own view of the march from downtown:

The majority of people that I was able to see in the crowd are mostly Latinos ... mostly people from Mexico. There are a lot of different Latin American countries, but it is mostly people of Hispanic origin. And they're here because they're upset about this proposal by the House of Representatives and they know there's this upcoming debate in the senate and they want to be heard. (Eliot and Kahn 2006)

When asked to speculate on the motivations of Latinx protesters, Kahn suggested that the demonstrators might be somewhat confused about the nuances of the legislation. But then Kahn went on to suggest that Latinxs were likely motivated by some sort of affront to their ethnic pride, stating: "They feel that they are the ones who are cleaning the office buildings, they are the ones that are taking care of the sick in this country, they're the ones that produce and they are not getting the recognition that they deserve."

Listening to Kahn's report, I thought of Schudson's critique of NPR, in which he argued that the network's lack of diversity in the newsroom will ultimately come through in its reporting. According to Schudson: "You know that most of these reporters will feel conspicuous and uncomfortable in many parts of their city. No doubt the staff makes an effort to cover issues of special importance to minorities and women, but you suspect that it is a mission and not a habit, and that it feels like a kind of foreign correspondence" (Schumacher-Matos 2012).

The *All Things Considered* segment makes evident NPR's lack of connection to the Latinx community, but it also reveals some of the limitations of balanced journalism on NPR. To present the state's argument, NPR featured the president of the United States, supported with the legitimacy of his office, and framing the discussion in his own voice. To capture the Latinx perspective, however, NPR turned to a non-Latinx journalist, reporting in English, and speculating on the motives of the participants, essentially rendering the Latinx community voiceless.

In the weeks that followed, NPR's coverage perpetuated Bush's dual frame of economic value and criminality. In addition to its rampant use of the term "illegal immigrant," NPR ran stories like, "Do illegal immigrants burden the justice system?" (Shapiro 2006) and "Q&A: Illegal Immigrants and the U.S. Economy" (Davidson 2006). In a segment recorded on March 30, NPR correspondent Ed Gordon interviewed Representative Maxine Waters about whether or not undocumented workers are taking jobs away from Black workers (Gordon 2006).

At other moments, NPR correspondents seemed blatantly insensitive to those Latinxs who had participated in the marches. During an interview on the Fox Network's *O'Reilly Factor*, NPR correspondent Juan Williams openly criticized young U.S. Latinxs for expressing their affinity with undocumented Mexicans, stating:

> These kids don't know anything. A lot of these are poor kids, struggling along in those schools and struggling to gain some sense of identity, so they're going to wave the Mexican flag because they feel somehow they are fighting for Mexicans living in the United States. And they're even going to get into crazy arguments about whether California should truly belong to the Mexico or the United States—all kinds of stupidity. (Maloy 2006)

NPR's coverage of the 2006 marches seems out of touch with the actual lived experiences of many Latinxs, especially when compared to the efforts of a number of commercial, Spanish-language radio stations at the time (Félix, González, and Ramirez 2008). In the days prior to the event, Spanish-language radio stations publicly promoted the rally, encouraging their listeners to protest in a peaceful and organized way (Block 2006; Flaccus 2006). These included Spanish-language radio personalities such as KLAX's El Cucuy en la Mañana and KSCA's Eddie "Piolín" Sotelo, whose morning talk shows were among the highest-rated programs in any language in Los Angeles (Watanabe and Becerra 2006b).

These efforts were consistent with Spanish-language radio's long-standing tradition of providing Latinx listeners with civic information and encouraging political participation (Casillas 2014). Spanish-language stations frequently help Latinx listeners to navigate the bureaucracies of U.S. institutions, often inviting experts such as doctors and lawyers to answer listener questions. In addition, Latinx DJs instruct their listeners on how to obtain immigration forms, fill them out, and register to vote (Baum 2006). Prior to the 2008 presidential elections, Spanish-language radio stations engaged in an effort to register a million new voters.

Despite these efforts, however, meaningful immigration reform failed to materialize in the aftermath of the protests. Instead, there have been a number of stunning reversals in recent years. During the 2016 elections, the GOP abandoned any pretenses of courting U.S. Latinxs, opting instead to engage in openly antagonistic rhetoric, backed by highly oppressive immigration policies, including the Trump administration's zero-tolerance policy, which has led to a number of human rights abuses, including the forced separation of families and the detainment of children at the border.

According to Amaya (2013), the limitations of Spanish-language media's organizing efforts highlight a paradox that defines Latinx political reality. Latinxs may have access to a robust Spanish-language media system, but they hold relatively little power to shape the political agenda. Furthermore, because the U.S. media landscape is highly fragmented by language, Latinx media is largely excluded from the national discourses that shape policy. Amaya further argues that English language media, which had increasingly focused its attention to the U.S.-Mexico border in the aftermath of 9/11, has contributed to the political rhetoric that has led to legislative measures targeting Latinxs.

Publics and Counter-Publics

In this book, I argue that NPR has, over the decades, secured its position as a white public space, while relegating Latinx listeners to the periphery. I further argue that NPR's positioning is tied to a larger cultural logic, in which Latinx identity is differentiated from national identity. It is a logic embedded in everyday broadcast practices and enacted by a range of public media practitioners, including station managers, news directors, producers, and editors. Finally, I argue that such practices have important implications for Latinx participation in civic discourses. After all, identifying who to include in one's imagined audience necessarily involves a process of who to exclude.

Certainly, the network did not start out this way. When it was originally created, NPR was designed with two clear mandates: to engage listeners in civic discourses, and to represent the diversity of the nation. During my research for this book, I had the opportunity to speak with Bill Siemering, who wrote the original mission statement for NPR in 1970. I asked Siemering about the network's original commitment to representing the entire nation and how that commitment was articulated publicly, to which he responded: "When I wrote [the statement], I was very keenly aware of 'Other' and the importance of inclusion. And that was the one thing that goes through that mission statement, which is inclusion and diversity and respect" (Siemering, interview with author, 2017).

According to McCauley (2005), NPR was born out of the desire to achieve a more democratic version of Habermas's public sphere, which critics saw as an idealized portrait of a bourgeois public sphere that favored propertied, educated males. While early critics of radio saw its broad reach as a defect, proponents of public radio saw its potential for achieving greater civic participation. It was precisely because of its ability to reach a wide audience that communities traditionally left out of civic discourses could now be included.

In his own critique of commercial media, Habermas (1970) was skeptical that radio could promote his notion of a public sphere. Because commercial media are driven primarily by economic interests, Habermas was doubtful that public discourse could occur outside the influence of the marketplace. However, advocates for public radio believed this issue could be addressed by its nonprofit status, thereby removing it the coercive influence of the state and the marketplace.

The Public Broadcasting Act, the legislation that established the Corporation for Public Broadcasting (CPB), gave the corporation responsibility to provide federal funds and political insulation for public broadcasting's two arms: NPR and Public Broadcasting Service (PBS). By separating NPR's funding from both the government and Corporate America, NPR would be able to pursue stories that better serve the interests of the public (Engelman 1996). "We are unbought and unbossed" is how NPR's Michel Martin put it, borrowing a quote from Shirley Chisholm: "Our listeners own us" (Holmes 2017).

NPR's funding model was meant to ensure that NPR would be free to reach a broader, more inclusive view of the American electorate, including listeners traditionally left out of civic discourses. In his original mission statement for the network, Siemering asserted NPR's mission of representing a broader cross section of the American electorate: "As man pulls himself out of the mass society to develop his unique humanness, his minority identification (ethnic, cultural, value) becomes increasingly important. This diversity is partially reflected in print media, but has not been manifested in the electronic" (Siemering 1970).

NPR's capacity to include minority producers would be facilitated by its tiered organizational structure. NPR's national overlay was seen as a way to serve the full range of American society, while NPR member stations dispersed throughout the country could reflect the nation's regional and cultural diversity. Over time, however, NPR's civic mission has been profoundly impacted by its economic reality. As McCourt (1999) has pointed out, the CPB, NPR, and PBS are all privately owned, not-for-profit organizations that are supported largely by private funding from corporations, foundations, and philanthropists. While privatization may partially insulate the state interests, it nevertheless subjects NPR to the marketplace, which in turn, has shaped the audience it pursues.

NPR may have originally conceived of its listeners as citizens bound together by national or political interest, but the network has, over time, increasingly defined its public as an audience that can be measured, to finally a market that can be exploited for underwriting purposes (Stavitsky 1995; McCourt 1999). Today, there appears to be significant slippage between the terms *the listener, the public,* and *the audience.* In response to economic pressures, NPR has created programming that would appeal to an educated audience with enough disposable income to make generous annual membership donations and to appeal to corporate underwriters (Loviglio 2013).

NPR's economic considerations, however, are not independent from its racial considerations. Despite its mandate to reach a broader public, NPR has consistently delivered programming to a narrow audience of educated, middle-class white listeners. By situating whiteness and privilege it its center, however, NPR has consistently moved minority audiences to the periphery. NPR's failure to achieve its diversity mandate has been well documented. By the mid-1970s, minority groups had already begun to express public concern regarding inequity in public broadcasting. In 1976, two hearings were conducted by the House Communications Subcommittee to address the hiring practices and professional norms that excluded minorities (Berkman 1980).

In response, the CPB decided to study the issue, and in 1977 they convened the Task Force on Minorities, whose mandate was to examine public media's success at addressing the needs of minority publics. The task force itself was diverse. Of its twenty-eight members, twelve were African American, seven were Latinx, three Asian American, three Native American, and three white. The task force met eight times in seven cities (Berkman 1980).

The product of this task force was a 1978 report titled *A Formula for Change: The Report on Minorities in Public Broadcasting*, which cited a number of shortcomings. While the report did not yield significant change at NPR or PBS, De La Cruz (2017) argues that the establishment of the Task Force on Minorities did result in the channeling of support and financing to minority and community radio stations. This, in turn, led to a flourishing Spanish-language public radio system, including Radio Bilingüe, a community-based radio network founded in 1976.

Radio Bilingüe may be seen as an example of what Fraser (1990) refers to as a *subaltern counter-public*, or an alternative discursive space that develops in parallel to official public spheres and "where members of subordinated social groups invent and circulate counter discourses to formulate oppositional interpretations of their identities, interests, and needs." From this perspective, Spanish-language public radio serves as a separate Latinx public sphere, where matters of Latinx public interest might be held outside white public space. This is not to say, however, that Latinxs have access to civic discourses that can shape political life. As Amaya (2013) argues, not all speech is considered equal, and within the majoritarian public sphere, English-language media is in a much greater position to shape the political agenda.

Rethinking the American Public

The current framework of two public radio systems, a Spanish-language one for Latinxs and an English-language one for "the nation," is an extension of the public sphere, which has long demarcated Latinxs from Americans. As Dávila (2008) argues, regardless of their residential tenure or legal status, Latinxs are continuously perceived as outsiders and "immigrants." These perceptions are based in the regulation and maintenance of a particular national identity, which elevates whiteness as its norm.

News organizations have, to varying degrees, been complicit in promoting these discourses. Within popular imagination, Latinxs are depicted as perpetual outsiders who threaten national security, what Leo Chávez (2008) calls "the Latino threat." U.S. Latinxs are said to be undermining national interests by overwhelming the population and weakening the country's key social institutions. It is argued that Latinxs are either unwilling or uninterested in assimilating, instead maintaining loyalty to their country of origin. They refuse to learn English. In its most base form, U.S. Latinxs are compared to drug traffickers, gang members, murderers, rapists, and terrorists.

While a number of actors are invested in perpetuating this discourse, news organizations have the capacity to wield a particularly pernicious form of symbolic violence. Journalists possess what Bourdieu (1993) calls the power of cultural consecration, meaning that they have the power to legitimize particular people and points of view. This is not to say that news organizations are acting maliciously when they circulate these narratives. After all, ideology is a system of thought that maintains particular forms of power relations that are often invisible to those who are actively involved in its production. Journalists may act in good faith, but they have their own presuppositions about the world, which inform their professional ideals and practices. Furthermore, through their agenda-setting and framing functions, news organizations act politically when they choose to report some stories while ignoring others. Or when they invest in the reporting of some communities while further marginalizing others.

NPR, however, was meant to serve as a unique kind of news organization. As a network that purports to represent the nation, NPR asserts unique claims about not only *what* it means to be American but also *who* gets to be American. In this way, NPR serves as an important cultural forum, in which listeners can imagine the nation and negotiate their places within it. This means defining the public in broader, more inclusive terms. As McCourt (1999) argues, NPR's strategy of pursuing small, homogenous populations runs contrary to the very idea of public radio, which implies public ownership and control. As a "public" broadcasting system, there is the expectation that NPR serve as space where diverse communities can come together to form shared identities and deliberate issues of concern.

Sound becomes essential for how NPR's listeners might imagine themselves as members of the American public. As Andrisani (2015) contends, it is through the simple act of listening that those without sustained access to political power might imagine alternative political possibilities, in which they are included. Here, Andrisani regards citizenship not as an institutionalized status that is bestowed by the nation-state, but rather as an informal social status co-constituted by members of a community. Nor is political participation limited to formal acts of engagement; rather, one's right to belong is asserted in the practices of everyday lived experience. From this perspective, sharing stories in Spanish, uttering words and phrases in accented English, or playing Reggaeton music are all expressions of citizenship that have strong sonic dimensions.

By providing a space where Latinx cultural producers might tell their own stories, in their own voices, NPR holds the promise of expanding notions of citizenship, in which Latinxs can assert their unique differences without giving up the right to be considered members of the public. In turn, through the shared act of listening, Latinxs listeners might negotiate, and even resist, dominant norms of what it means to belong, enacting what Anguiano (2018) refers to as "sonic citizenship." Sonic elements such as voice and music are important for helping Latinx listeners to understand citizenship on their own terms.

Rather than appeal to a broader definition of citizenship, however, NPR has more often reified existing social hierarchies. On air, NPR remains a space where primarily white voices can be heard. A 2018 study of the racial and ethnic diversity of sources on NPR's weekday newsmagazines revealed that white voices made up 83 percent of sources heard on NPR, while Latinx voices accounted for only 6 percent (Jensen 2019a). The use of sound to exclude, rather than include, also has a longstanding history. As Western (2020) argues, exclusionary practices have often been sonic ones, and simply sounding foreign is enough to exclude one from the political center.

This conception of the American public, however, is becoming increasingly at odds with a public that is becoming more ethnically, racially, and linguistically diverse. In recent years, the country has experienced profound demographic change, raising new questions regarding how media organizations create programming for the audiences they purport to serve. Latinxs have figured prominently in this change. At 18 percent of the country's population (U.S. Census 2020), U.S. Latinxs are becoming increasingly important to the American electorate that NPR is tasked with serving. This number is expected to grow exponentially in the next few years. By 2045, Latinxs will account for almost 25 percent of the country's population (Frey 2018). Furthermore, the population is defined by its youth. About one-third of U.S. Latinxs are under the age of eighteen, which means that they figure prominently in NPR's audience of the future (Patten 2016).

Since 2000, the primary source of Latinx population growth has swung from immigration to native births, which runs contrary to narrative of Latinxs as perpetual immigrants, without the capacity to engage in civic practices. Between 2000 and 2010, there were 9.6 million Latinx births in the United States. Overall, U.S. births alone accounted for 60 percent of Latinx population growth (Krogstad and López 2014). Between 2010 and 2015, there were 5 million Latinx births in the United States, compared with just 1.9 million newly arrived Latinx immigrants (Flores 2017).

According to the Pew Research Center (Krogstad, López, López, Passel, and Patten 2016), the size of the Latinx electorate was expected to number 27.3 million eligible voters in 2016 and is projected to make up 12 percent of all eligible voters, a share equal to that of African American eligible voters. Despite this growth, Latinxs have not engaged politically. The Pew Research Center also found that voter turnout lagged among Latinxs. In the 2016 presidential election, Latinxs accounted for only 9 percent of voters, compared to 12 percent for African Americans, and 73 percent for whites (File 2018).

Associational, linguistic, occupational, and residential factors have historically contributed to placing Latinxs in a separate social sphere (Massey and Denton 1993). This kind of isolation, in turn, has excluded them from the civic discourses that impact local and national interests. Given the paucity of civic information circulating amongst Latinxs, NPR has the potential to serve as an important change agent. With over one thousand member stations, NPR has the capacity to reach 98.5 percent of the country's listening audience (National Public Radio 2020a). In 2017, 37.7 million listeners tuned into NPR stations every week (National Public Radio 2018). With seventeen domestic bureaus and another seventeen international bureaus, NPR is in the position not only to shape the public agenda, but also to frame specific policy issues. Several of NPR's domestic bureaus are based in areas that are heavily populated by Latinxs, including Los Angeles, New York, Chicago, and Miami. NPR's presence in these communities gives their correspondents access to Latinx stories, but also places them in a position to best understand the civic issues most relevant to the Latinx community.

Despite this promise, NPR has struggled to serve the Latinx listener. According to a 2019 report, Latinxs accounted for just over 6 percent of the newsroom staff, prompting Keith Woods, NPR's vice president of newsroom training and diversity, to admit that the hiring of Latinxs remains a "chronic need" (Jensen 2019b). A year earlier, Christopher Turpin, vice president for Special Projects+Innovation at NPR, put it more pointedly, stating "I think we have done a very bad job around Latinxs and diversity. I think it's an area where our coverage really suffers from the lack of people who have insight into a community" (Jensen 2018).

These practices are not sustainable. The baby boomers who NPR has long pursued are now reaching sixty-five and older, which means that NPR is finding itself in a position where its audience is becoming older and whiter, while the population at large skews younger and more ethnically diverse. According to the public-media website *Current*, the audience for *Morning Edition*, NPR's most popular show, has fallen 11 percent since 2010. Among listeners under the age of fifty-five, listenership has declined by 20 percent. The one group that's listening more is those sixty-five and over. NPR projects that, by 2020, the number of listeners forty-four years old and younger who listen to NPR stations will be around 30 percent, half of what it was in 1985 (Falk 2015a). In 1995, the median age of the NPR listener was forty-five years old. Today, the average listener is fifty-four (Farhi 2015).

NPR is not unaware of their problem addressing Latinxs. In articulating their strategic vision for the future, NPR acknowledged that they need to create or showcase talent that appeals to people of varied ages, ideologies, and ethnic backgrounds, with an emphasis on Latinx audiences as one of the fastest-growing demographic segments, using relevant platforms and venues. In 2016, Joseph Tovares, CPB's vice-president of diversity and innovation, described these imperatives: "I decided that the best argument to advance diversity in the system was the 'diversity as a business imperative model.' That remains a great and pragmatic argument. But six years later, I also see diversity as a moral imperative—it is simply the right thing to do . . . it's more than just the bottom line" (Tovares 2016).

Here, Tovares argues that NPR's civic mission must extend beyond its economic goals. However, this ideal has been hard to achieve, given the competitive landscape in which market pressures are formidable. During a moment of introspection, NPR invited a number of national leaders to publicly reflect on NPR's diversity efforts (Schumacher-Matos 2012). In the piece, Janet Murguía, president and CEO of National Council of La Raza, directly addressed NPR's challenge engaging Latinx listeners:

> Should NPR focus on appealing to its core audience, as most commercial broadcasters do, or does it have a broader mission to serve its public, given its unique role in America? I would argue that focusing on just serving college graduates underestimates the appeal and value of NPR's programming to the Latino community.

According to Murguía, NPR must privilege its civic mission over its economic needs. If NPR is committed to growing its Latinx audience, it must invest in the resources needed to hire more Latinxs as journalists and producers and to invest in new programming. It must also be willing to diverge from its current strategy of reaching affluent, educated audiences, who have historically supported NPR member stations.

To better understand the practical decisions that shape newsroom practices, it is important to consider NPR in its broader context. Like every event and every social institution, NPR must be seen as a historical product that has evolved in concert with significant changes in technology, industry practice, and demography. To ascertain the degree to which NPR has deviated from its original mission, it is important to ask the following questions: What kind of news organization was NPR intended to be? What has it become over time? And in what ways is it evolving to meet the needs of a nation that is becoming more ethnically, culturally, and linguistically diverse?

To examine these questions, I draw upon previous research conducted by scholars working in both Latinx Media Studies and public radio. Much has been written about National Public Radio and its role in American society. Certainly, NPR has curated its own legacy, producing a vast number of cultural texts that articulate its perspective on a variety of topics including music (Libbey 2006; Schoenberg 2002), politics and culture (Stamberg 2012), and broadcast production (Kern 2008), as well as its own institutional history. In its retrospective, *This is NPR: The First Forty Years* (2010), Cokie Roberts, Susan Stamberg, and other NPR personalities nostalgically recount the network's growth over the first four decades—from a news staff of just five reporters to a global news organization with domestic and international bureaus.

In this retrospective, NPR maintains that it is has retained its journalistic ethos, but scholars have painted a more complicated picture of the network's evolution over time. For example, Engelman (1996) and McCauley (2005) have argued that what started as an insurgent, quasi-amateurish operation has evolved into a highly professionalized organization with tremendous political and social influence. Others have been more critical. Scholars, including Loviglio (2013), McChesney (2008), McCourt (1999), and Stavitsky (1995) have separately argued that, in an effort to secure stable funding, NPR has adopted a neoliberal sensibility, which has undermined its ability to serve the public interest.

Much of the research on NPR has focused on the network's economic pressures, but there has been little research directly addressing the issue of NPR's diversity mandate. Furthermore, a focus on how NPR has served the Latinx listener over its fifty-year history has been left relatively unexplored. To remedy this, I engage the literature on Latinx media studies, with a particular focus on those scholars working on Latinx radio. Scholars including Gutiérrez and Schement (1979) and Castañeda (2008) have chronicled the evolution of Latinx radio from hourly programming restricted to undesirable dayparts into a thriving system with transnational connections. Over time, Latinx radio has become an important space for working-class, Spanish-dominant Latinxs to negotiate their place within the dominant culture (Anguiano, 2018; Casillas, 2014). In her book, *Sounds of*

Belonging (2014), Casillas argues that these stations affirmed the ethnic identities of their listeners and promoted direct civic participation, which included organizing protests, informing listeners about policy, and promoting voter drives.

Focusing specifically on public radio stations, De La Cruz (2017) has documented the ways in which public radio stations have successfully engaged Latinxs civically and socially. Castañeda (2016) sees the role of community radio as particularly important at a moment in which Latinxs are particularly embattled, proposing that community radio stations might partner with social justice organizations that serve local communities, while also engaging with national issues that affect Latinxs. With such partnerships, radio stations can inform communities about local healthcare, ICE work raids, and important strategies for keeping Latinx families safe.

Researchers have also examined the ways in which Latinx public radio serves as a form of language preservation. Focusing on the case of *Radio Indígena*, an immigrant-owned-and-operated community radio station in Oxnard, California, Jiménez found that the station provides essential health information and humanitarian support in several indigenous languages, including Mixteco, Zapoteco, and Triqui. Stations that blend Spanish, English, and Nahuatl are unique in that they resist the notion of monolingualism as their starting point. This fluid conceptualization of language and linguistic practice differs significantly from that of NPR, in which stories about Latinxs are almost exclusively told in English. As Noel (2017) found, when Spanish is included, it is often rendered as foreign. Marked as "ethnic noise" in NPR transcripts, Spanish is often treated as an ambient sound, unworthy of translation.

Each of these scholars highlights the importance of multilingual, Latinx-oriented public radio as an important resource. At the same time, the framing of Latinx public radio as separate from "national public radio" reflects a media landscape that has been organized on the basis of linguistic usage and ethnic identity. By reclaiming NPR, I extend Amaya's (2013) argument that continued support for a distinct Latinx public sphere must be balanced with the need to participate in the majoritarian public sphere. Without this access, Latinxs are constrained in their ability to engage civically or to shape policies that impact our communities.

The genesis of this project came during an unexpected moment, when I was attending a meeting at NPR member station KLCC in Eugene, Oregon. One day, during the spring of 2016, we had learned that Donald Trump, then a candidate, was preparing to hold a campaign rally in our city. This prompted a lot of speculation within the newsroom. After all, why would a conservative candidate hold a rally in a college town in a dependably blue state? As we debated the issue, Tripp Sommer, the station's news director, said casually on his way out the door, "well, we're going to cover this candidate like we'd cover any other candidate."

What Mr. Sommer said was not necessarily controversial. After all, Sommer was simply articulating the journalistic perspective. As Darras (2005) argues, journalists carry the doxic belief that they have a democratic duty to cover people of importance, with particular priority given to presidential candidates, who are seen as future decision makers. This belief is itself tied to a larger assumption that only the government possesses political power and is, therefore, worthy of press coverage.

But I couldn't help but think that these journalistic practices obscured more than they revealed. After all, Mr. Trump was anything but a normal candidate. By the time that Trump had come to Eugene, he had already built a viable candidacy largely by attacking Latinxs, citizens and noncitizens alike. Instead of addressing these concerns directly, the station ran a story that seemed to blur the lines between journalism and public relations. Online, the story ran with the headline "Trump Supporter Says Eugene Rally Was 'Electrifying'" (Buckley 2016). In it, journalist Kyra Buckley made sure to establish her journalistic objectivity by interviewing Karen Clark, a voter who claimed to be neither Democratic nor Republican. "I really have a lot of respect for Mr. Trump since watching his show," Clark says during the interview, "and I saw how much common sense he had and how fair he was with people."

While journalists may see Trump's rally as news, the event may be better understood as what Boorstin (1971) refers to as a pseudo-event, one produced solely for the purpose of gathering media attention. By covering the event uncritically, NPR journalists became complicit in what Champagne (2005) describes as their own passive manipulation by politicians and their handlers, who engage in constant efforts of mystification aimed at journalists through press files and organized visits.

These practices run contrary to the very notion of public radio, which was meant to operate outside the influence of both the state and the marketplace. NPR's architects believed that they could remain independent of the forms of censorship that inhibited commercial news organizations from truly serving the public interest. In theory, NPR would have more control over the content they produced. They could also have freedom of topics to choose from and could implement time limits that would allow them to address those topics critically, and in detail. Rather than pursuing consumers for advertised products, NPR would also have the freedom to reach various "publics."

I spoke with Frank Cruz, former chair of the CPB, about NPR's struggle to reach Latinx listeners. Cruz told me that part of the issue is funding. Because its financial position has always been tentative, the network has been motivated to seek out audiences that could financially support it. According to Cruz:

> When it was created in 1967 by Lyndon Johnson, and they created the Public Broadcasting Act. What they should have done then. They should have provided a good financial foundation. They didn't. What it did, it left public broadcasting—CPB, and then PBS the year after that and NPR the year after—it left us in the position of having to go to on bended knee to Congress to ask for funding. That was one difficult stroke right there. (Cruz, interview with author, 2020)

To better understand the economic factors that shape the ways in which NPR enacts its diversity mandate, I employ a critical political economy approach, which calls for attention to the interplay between symbolic and economic dimensions of the production of meaning (Hardy 2014). From this perspective, it means looking at NPR as both a symbolic and an economic product. On one hand, research in this area must attend to the way in which various media organizations are funded, organized, and regulated (Golding and Murdock 1991). On

Introduction

the other hand, there is a need to understand how ideology is embedded within media organizations.

According to Hardy (2014), there are three main implications for critical political economy research. First, Hardy argues that researchers must provide a careful study of how a given media organization works, including ownership and finance. For example, understanding NPR means understanding the ways in which the policies and actions of the state affect NPR's media behavior and content. Despite its public-funding model, NPR is still beholden to the federal government, which regulates its content and provides the annual subsidies, albeit limited, to support its operations. Even as I was writing this book, President Trump had publicly threatened to defund NPR, prompting some legislators to intervene on NPR's behalf (Lowey 2020). To buffer itself from a capricious political landscape, the network has increasingly subjected itself to listener-subscribers who provide the annual donations for NPR member stations as well as the organizations that provide the corporate underwriting dollars (McChesney 2008).

Second, researchers must attend to different ways of organizing the media as well as the changing nature of the industry's structural influences. From a research perspective, this means taking a critical look at the naturalized, taken-for-granted positions that the define the news industry, such as *public radio* and *commercial radio*. The relational perspective forces the researcher to ask some questions, such as "what does it mean to be a public station?" and "public in relation to what?" It is the relational perspective that allows researchers to draw a map of where NPR is placed alongside other news organizations. Because the specific focus of this book is on the Latinx community, this also means taking a look at the differences between *national news* and *ethnic* or *Hispanic news*, which itself reflects a system of categorization that distinguishes Latinx identity from national identity.

To this end, Hardy argues that researchers must attend to the relationship between media and the broader structure of society. To understand any media, investigators must address how it is produced and distributed in a given society and how it is situated in the dominant social structure (Kellner 2009). A main critique of political economy research is its primary focus on economic influences without adequate consideration for non-commercial ideologies. I address this deficit by understanding how the dynamics within NPR are inextricably linked to larger questions of power. In doing so, I engage research from a Latinx media studies tradition, which attempts to understand the role of Latinxs as both participants in, and the subjects of, media industry practices. As with other forms of media studies, research in this area looks at the triadic relationship between producer, text, and audience, but Latinx media studies scholars have generally integrated themes of nationalism, citizenship, and immigration, as well as the unique roles of language, race, class, and gender (Valdivia 2010).

I want to start by acknowledging that NPR is a large, multifaceted media organization. While it is largely governed by decision makers working at its headquarters in Washington, D.C., NPR works with local member stations throughout the country that are managed somewhat independently. Furthermore, NPR collaborates with a number of independent producers who are, themselves, independently governed, yet, at the same time, subject to pressures exerted by NPR. Because programming has, over time, become more centralized, I focus primarily on decision-making processes at the national level, yet I am attentive to how these processes impact NPR member stations and independent producers.

The data generated during this investigation primarily focused on institutional discourses, which included both internal and external documents. Internal data included NPR-generated communications intended for a variety of publics, including its listeners, prospective corporate sponsors and private donors, and regulators, as well as its own community of stations and employees. I also examined NPR's branded material (books, music, merchandise), which is meant to establish a relationship between the network and its listener-subscribers. External discourses included syndicated research (both sponsored and governmental), NGO reports, and government documents, as well as trade press regarding National Public Radio and other noncommercial broadcasting entities.

Industry data was supplemented with more than fifty in-depth interviews conducted with public radio practitioners. These included journalists, hosts, producers, public editors, station presidents, general managers, news directors, members of the Corporation for Public Broadcasting, and independent producers who have worked with NPR at the national and local levels. Because some of the participants in this research are public figures, I have decided to include their names. When the identities of participants are not relevant, however, I have not identified them by name.

Qualitative interviews are an especially useful method for understanding industrial practices, because they are designed to take the investigator into the lifeworld of the informant and to see the content and pattern of daily experience (McCracken 1988; Mason 2002). Furthermore, qualitative interviewing emphasizes depth, nuance, complexity, and roundedness of data, as opposed to other methodological approaches. The basic assumption of qualitative research is that, by spending enough time with the participant, the investigator is taken backstage to the culture in question and provided a glimpse into the assumptions and categories that are otherwise hidden from view. Because this project employs concepts that may not be readily accessible to participants, qualitative interviews afford the researcher the flexibility to probe the particular meanings elicited in informants' testimonies.

With regard to my own positionality, I want to disclose that I am not a journalist—a reality that allows me to see some things, but not see others. While I do not teach newsroom practices, many of the students I advise and teach go on to work in journalism, and some of them have started their careers by working at NPR and their member stations. They learn invaluable skills in these internships, including how to produce "quality journalism." But I also bear witness to other forms of transformation. I find that these students adopt the sensibilities that will make them suitable for the newsroom, and they learn to modulate their voices to be appropriate for broadcasting purposes. In Bourdieu's (1991) terms, they are being inculcated into their profession. By the time these students graduate, their dispositions, behaviors, and ways of thinking, writing, and speaking are, to varying degrees, perfectly suited to their future careers in journalism. They will become members of a field that replicates itself.

But as a Latinx media scholar, I cannot help but think that these practices lead to homogeneity at a moment when our society needs media diversity. My dual role as consumer and critic of NPR allowed me to interrogate different perspectives. During my conversations with NPR practitioners, we were able to discuss specific industry practices that ultimately predetermine the stories we hear. However, I can only assume that my identity as a Latinx scholar facilitated some conversations, while impeding others. A Latinx researcher interrogating issues about the ways in which NPR enacts its diversity mandate has the potential to make practitioners apprehensive, yet participants were generally forthcoming in their responses. I was particularly appreciative of candid testimony about NPR's racializing practices, especially from Latinx practitioners who were reflecting on their place in a profession in which they are a clear minority.

Roadmap to This Book

The book is organized into two main sections. In the first half of the book, I examine the specific logics of practice that undermine NPR's capacity to serve the Latinx community. In the first chapter, I examine how NPR imagines its ideal Latinx listener. In doing so, I trace NPR's evolution from thinking of its listeners as a public, to thinking of them as an audience, and finally to thinking of them as a market that can be segmented for underwriting purposes. Here, I draw from Ang's (1991) argument that the audience is a rhetorical construction that is created by, and in the interest of, the institution. From this perspective, network executives do not simply tap into a community that is, objectively, "already there." Instead, media institutions shape the contours of those particular groups in order to present a credible, desirable, and marketable audience. Choosing a target has practical and political implications. Audience construction does more than merely set the preconditions for what the programming priorities will be. Choosing a target involves the process of identifying who to engage in civic discourses and who to exclude.

In chapter 2, I focus on the NPR voice and the standard language ideologies of public radio. As a primarily audio-driven medium, NPR's use of voice raises unique considerations about the nature of on-air diversity. Drawing primarily on the field of sociolinguistics, I argue that NPR has cultivated an idealized dialect, which situates whiteness at its center. NPR has continued the long-standing practice within the broadcasting industry of employing Standard American English (SAE), in which regionally and ethnically marked features have been suppressed, so that they are commonly perceived as neutral. I argue that NPR employs standardizing practices that privilege white, educated, middle-class listeners, while relegating Latinx speakers to the periphery. Grounding my analysis in the concept of standard language ideology, I argue that such standardizing practices serve as a form of language-based discrimination, excluding speakers of traditionally stigmatized varieties from political discourses.

In the second half of the book, I reverse the angle and examine NPR from the point of view of Latinx cultural producers who are producing shows that address Latinx issues. In this section, I present three case studies of national, Latinx-oriented programs that are affiliated with NPR. Here, I am interested in the ways in which Latinx practitioners, working with NPR, negotiate—and, at times, subvert—the network's institutional norms. In doing so, I draw primarily from the field of critical media industry studies (Havens, Lotz and Tinic 2009), which accounts for the agency of individual cultural producers who often engage in practices and competing goals, which are not subject to direct and regular oversight by institutions.

In chapter 3, I focus on *Latino USA*, a syndicated radio program produced by María Hinojosa's Futuro Media Group. *Latino USA* was first designed for terrestrial distribution and focuses on national issues. But what started as an upstart radio program has evolved into a multimedia destination, which includes a strong online component. Furthermore, Hinojosa and her team have been successful at building partnerships with both commercial and noncommercial entities. Based on interviews with Hinojosa and her team in Harlem, New York, I examine the ways in which *Latino USA* creates news that is designed to address significant Latinx issues. Furthermore, I examine the ways in which Hinojosa and her team negotiated the power-laden relationship with NPR, who served as the primary distributor of the program until 2020.

In chapter 4, I examine *Radio Ambulante*, NPR's only Spanish-language podcast. While *Latino USA* and *Radio Ambulante* are intended, to varying degrees, to engage Latinx listeners, each program conceptualizes the Latinx listener differently, and each is grounded in a different platform that shapes its intended audience. Unlike *Latino USA*, which was conceived of as a terrestrial program, *Radio Ambulante* was conceived of as an entirely digital enterprise. Furthermore, its ideal Latinx listener is one with a more global sensibility, part of an imagined community that spans the Americas. Consequently, its producers have benefitted from the transnational flows characteristic of globalization, where online networks have replaced traditional geographical spaces as the primary modes of affiliation. This has endowed a relatively small team of journalists and producers with access to the means of production, with the ability to reach listeners beyond the boundaries of the United States, and with the capacity to build new kinds of collaborations with journalists based throughout the Americas.

In chapter 5, I focus on NPR's *Alt.Latino*, a music-oriented podcast that showcases Latinx alternative artists while addressing current political, social, and cultural issues. Established by Felix Contreras and Jasmine Garsd, the podcast has the potential to serve as a space by Latinxs for Latinxs. However, I examine how *Alt.Latino*'s ability to address important social issues relevant to Latinx listeners is constrained by the network's institutional norms and practices. Based on a review of NPR discourses, public statements by the show's producers, and interviews with the show's creators, I argue that *Alt.Latino*'s designation as a "music" show enables it to engage in oppositional work not possible in NPR's general news programming. But I further argue that, as an NPR program, *Alt.Latino* is beholden to the network's editorial guidelines, thereby impeding its mission to reach out to disenfranchised listeners.

In the concluding chapter, I revisit NPR's mission to civically engage its listeners, a mandate that was embedded within the network's original design. Over the course of its history, NPR has cultivated an ethic of objective, passive journalism, an ethos that has been codified in its institutional practices and guidelines. As stated in NPR's ethics guidelines, "we have opinions, like all people. But the public deserves factual reporting and informed analysis without our opinions influencing what they hear or see" (National Public Radio, 2019). NPR's rhetoric of impartiality takes a clear moral stand, but it is one that ultimately obscures the ideological work conducted by the network. After all, it is the publicly funded nature of NPR that gives the network the pretense of objectivity, while preserving existing power structures.

Here, I attempt to situate NPR in relation to other news organizations, which communicate a greater diversity of information and opinion, feature a wider array of ideologies, and do more to promote political participation. As part of this chapter, I also explore the possibilities within the current public radio framework to reinsert Latinxs and Latinx perspectives into NPR. During my research, I encountered a number of tactics meant to address this issue. None of these rec-ommendations were new. All came up, in various forms, during my conversations with practitioners at the local and national level.

By exploring these issues, I hope to address how power is en-acted in everyday broadcast practices, which is the ultimate focus of this book. By interrogating industry practices, we might begin to reimagine NPR as a public good that is meant to be accessed by the broader spectrum of the American public, not just the country's most elite. Furthermore, a focus on industry norms and practices allows us to engage the much larger conversation around what it means to be American. Finally, by diminishing the barriers that separate Latinx media from national media, we might also begin to rethink public radio's civic role in ways that extend beyond dispassionate forms of journalism to include more direct forms of political participation.

NPR'S PURSUIT OF THE IDEAL LATINX LISTENER

Programming is a lot like bait. What we catch depends on what we set out. Honey draws bees, worms lure fish, and a hunk of liver will bring stray cats to your door. . . . In the same way, certain kinds of listeners are attracted to certain kinds of programming. So, when we choose what we air, we select who will listen—and also who won't.*

— DAVID GIOVANNONI, NPR RESEARCHER

In 1988, NPR announced its decision to cancel *Enfoque Nacional*, a thirty-minute Spanish-language news magazine program that was produced out of NPR member station KPBS-FM, San Diego. The show ran for nine years, focusing on national issues of concern to Latinxs. It also served as a training ground for Latinx journalists in public radio, including NPR education correspondent Claudio Sánchez, and María Hinojosa, who later went on to headline *Latino USA*.

In an effort to justify the decision, NPR executives cited a decline in listeners and lack of member-station interest as the reasons for the show's demise. At the time, only 6 percent of NPR member stations subscribed to the program. Furthermore, with a weekly audience of eleven thousand, NPR executives argued that *Enfoque Nacional* did not warrant an annual budget of $180,000. However, critics of the move believed that NPR never truly supported the program. Arnold Torres, a Sacramento lobbyist on Latinx issues, argued that NPR had abdicated its diversity mandate by focusing only on the nation's most privileged listeners, stating "NPR has always ghettoized minority programming. The problem is [that] public radio was not designed to simply have white liberals listen to what they want to listen to. It is designed to provide alternative programming" (Harper 1988a).

---

*From Audience 98: Public Service, Public Support, 7.

During an interview with the *Los Angeles Times* (1988), NPR's president at the time, Douglas J. Bennet, seemed to suggest that the network was neither motivated nor capable of reaching Latinx listeners in Spanish. In what seemed to be a circular argument, Bennet claimed that NPR should not invest in Spanish-language programming because the network did not reach Spanish-speaking listeners. "My argument from the beginning has been to go to commercial stations," Bennet stated during the interview. "The point is to reach the Spanish-speaking audiences" (Harper 1988b).

Bennet's comments are remarkable for a number of reasons. First, Bennet provides insight into the direct link between audience research and programming decisions in public radio. Using marketplace logic, Bennet claims that, if programs are to be funded and supported by NPR, they must deliver audiences that NPR finds appealing, and in sufficient amounts. Perhaps more remarkable is Bennet's claim that Latinxs would be better served by commercial radio, which is a stunning reversal from the network's long-standing position that commercial radio stations are incapable of serving the public interest because they are beholden to the interests of advertisers. Rather than framing NPR as an alternative to commercial radio stations, Bennet is affirming the notion that NPR is just another a niche product within the larger broadcasting enterprise.

By 1988, NPR seemed to have taken a clear position that Spanish-speaking Latinxs were not considered part of the public that NPR was tasked with "luring" with dedicated programming. Yet, over its fifty-year history, the network has continued to benefit from the ongoing claim that it serves the public. This begs the question: How exactly do U.S. Latinxs figure into NPR's imagination of the American public?

In this chapter, I trace NPR's evolution from thinking of the listener as a member of the American public, to constructing them as an audience that can be measured for programming purposes, to ultimately thinking of them as a market that can be exploited for economic advantage. I begin by highlighting the distinction between NPR's rhetoric of inclusivity and its actual broadcasting practices, which have ensured that it speaks to a small but privileged segment of the listening audience. Drawing heavily from commercial practices, NPR has engaged in an ongoing practice of market segmentation, in which the public has been divided into subgroups based on some shared traits in an effort to efficiently channel resources (Tynan and Drayton 2020). I further argue that this practice has fundamentally shaped NPR's diversity mandate. By segmenting the Latinx audience, NPR has chosen to focus on those listeners who are most congruent with their target profile, resulting in a conservative, rather than transformative, programming strategy.

A Precarious Funding Model

When *Morning Edition* first launched in 1979, NPR producer Rick Lewis appealed to member stations to support the integration of the program system-wide. In doing so, Lewis invoked a broad definition of the term *public*, stating "we try to mirror ALL of the country—perhaps the hardest thing of all. And we need the help of our member stations" (Roberts 2010). Over its five-decade history, NPR's rhetoric of serving the "public interest" has benefitted the organization, distinguishing it in a competitive media marketplace by investing it with a moral authority not available to commercial stations. "The mission of NPR is to work in partnership with Member Stations to create a more informed public" is how NPR currently promotes itself. "One challenged and invigorated by a deeper understanding and appreciation of events, ideas and cultures" (National Public Radio n.d.).

It is perhaps more accurate to describe NPR as a vehicle for distributing news-oriented programming designed for a narrow segment of the listening audience. The financial constraints placed on public media have required that NPR make specific choices about where to invest its resources. As Walker (2017) points out, if CPB dispensed funds to every small community, it would have to divide its budget so finely that no station would receive enough money to justify the corporation's existence. But as Ang (1991) makes clear, the "audience" is a rhetorical construction that is created by and in the interest of media organizations. Despite their claims of creative excellence and social importance, media institutions are designed first and foremost to pursue audiences that are essential to their economic well-being.

Discourses around audiences have generally focused on commercial media. In an effort to deliver consumers to marketers, media institutions are said to shape the contours of particular communities in order to present a credible, desirable audience. However, public radio raises unique considerations. As a state-supported effort, public radio is not quite a commercial enterprise. That is not to say, however, that NPR is not subject to economic pressures. Since its inception, NPR's funding model has been precarious. Originally, educational radio stations, the precursor to the NPR system, were supported by local institutions such as churches, unions, and universities (Loviglio 2013). With limited funding, these stations operated within constrained budgets, but when the1967 Public Broadcasting Act was passed, the goal was to provide sustained support for public radio.

At the time, NPR's funding drew annual grants allocated by the CPB, which meant that NPR had to operate with minimal funding. Consequently, NPR's first journalists received little compensation and had access to poor production equipment. During my interview with Jeff Kamen, one of the original journalists at NPR, we discussed the highly constrained conditions of working at NPR in those early days. While he appreciated the network's vision, Kamen described his time at NPR as a struggle to balance his idealism with the practical reality of working at a station with limited resources. Prior to his time at NPR, Kamen had gained substantial experience covering the civil rights movement, but also recognized he was hired because of his ability to work on a shoestring budget. As he put it, "I was hired because I was a one-man band. And I would go out, get the story, bring it home, cut it, and you could put it on the air" (Kamen, interview with author, 2019). Ultimately, Kamen's tenure at NPR was short-lived. The reality of working under such austere conditions was constraining. "Remember, I was only in Washington, D.C., for six weeks. I quit because they required me to fill out a form for getting new batteries for my field recorder," Kamen told me. "What insanity was that?"

It became apparent that, to succeed, NPR needed to become economically viable. According to McCourt (1999), NPR's funding model changed in 1985. Instead of funding NPR directly, the CPB would allocate funds to NPR's member stations across the country. Those stations would, in turn, use those funds to purchase programming from NPR as well as from other producers and distributors of public radio programming. This funding model freed up the network to pursue other forms of revenue: corporate underwriting and foundational support (Loviglio 2013).

Today, NPR relies on a mix of fees, grants, corporate donations, and federal funding. According to NPR's report on Public Radio Finances (2020b), dues and fees paid by NPR member stations are the largest portion of NPR's revenue (35 percent). The most significant component of station dues and fees revenue is the charge for carrying NPR's flagship newsmagazines—*Morning Edition, All Things Considered,* and *Weekend Edition.* These programs are priced based on the amount of listening per station multiplied by a common unit price. NPR charges more as the volume of listening increases, and, in turn, stations are able to raise more funds from their supporters and communities. Member stations' annual dues and fees for digital services make up the balance of stations dues and fees revenue.

Corporate donations account for 33 percent of the network's revenue sources. National Public Media (NPM), who serves as the exclusive sponsorship representative for NPR, actively promotes opportunities with corporate partners. These opportunities extend across NPR's platforms. In exchange for these donations, messages acknowledging NPR's national sponsors are presented on-air in short announcements and are presented in visual and audio form on NPR.org and through other digital services.

While corporate sponsors may be motivated by a sense of civic duty, the decision to fund NPR is also a transactional decision. Like other forms of marketing communications, corporate underwriters are seeking to reach an audience that cannot easily be reached through other media. Therefore, there must be a perceived fit between programming content and the tastes of their intended audience. However, when justifying the contribution of corporate underwriting to their overall revenue stream, NPR makes clear to the public that there is no relation between financing and editorial. "Corporate sponsors cannot influence NPR's coverage. NPR journalists have no role in selecting corporate sponsors," NPR states in its public disclosure of public radio finances (National Public Radio 2020b). "Our journalists are trained in the ethics and practices of journalism which prevent outside groups from influencing their objectivity, story selection, and reporting. . . . When news warrants, we will report on the activities of companies that support NPR."

At the member-station level, NPR also relies heavily on donations from individual listener-members, which, as Loviglio (2013) argues, is a way of conceiving of the listening audience that is consistent with a neo-liberal society. Rather than focusing on listeners as active members of the public, listeners are valued primarily for their economic functions.

The integration of marketplace logic into NPR's broadcasting practices has been attributed to David Giovannoni, the president of Audience Research Analysis (ARA), a consultancy that held contracts with the CPB, NPR, Public Radio International (PRI), and almost every major NPR member station in the country. Described by the *New York Times* as "public radio's private guru" (Freedman 2001), Giovannoni is credited for a prompting a shift from conceiving of the listener as a member of the public, to conceiving of them as an audience that can be measured for underwriting purposes. By providing data on who listens to NPR and how much, Giovannoni methodically assessed which listeners were most likely to donate. But while Giovannoni has been credited for helping NPR achieve more self-sufficiency, critics have argued that Giovannoni's methods have undermined the kind of programming that NPR was designed to produce in the first place.

From Public, to Audience, to Market

Just prior to the passing of the Public Broadcasting Act of 1967, much of the support for public broadcasting had focused on public television. To ensure that radio was not left out of any pending legislation, National Education Radio, a division of the National Association of Education Broadcasters, hired Herman W. Land Associates to study the field and its potential. The product of that study was *The Hidden Medium: A Status Report on Educational Radio in the United States* (1967), which was an assessment of public radio that would later provide the framework on which NPR would be established. At the time, the authors found that there was a virtual absence of research amongst public radio stations and that stations knew very little about their audiences. The committee found that over 50 percent of the stations conducted no audience research of any kind, and only about one-third attempted to determine the size of the audience. For example, 42 percent of the respondents asked to provide an estimate of the number of homes in their areas were unable to do so (Herman W. Land Associates, Inc. 1967).

According to the report, there appeared to be two reasons for the scarcity of research. First, the commission found that there was a lack of conviction amongst station managers that audience research was either necessary or important, given the cultural orientation of many stations. The second issue was budget. Research is expensive, and stations were ill-equipped to conduct even minimal studies. Nor did they have the resources to purchase third-party data. The result was that stations had little insight about the people who listened to their programs.

In the *Hidden Medium*, there is some insight into the nature and listening behavior of the audience for educational radio at the time. For example, they found that the audience for educational radio was not unlike the NPR listener of today. The listeners were generally above average in education and income and tended to be older than the average audiences listening to the popular music-and-news stations. Although the listener expressed some interest in other types of programming, music was the preferred format, particularly classical and lighter music. News was important, but not paramount. Listening was greatest in the evening hours, rising from 7 to 9 and "peaking" at 10 p.m., which reversed the pattern of commercial radio, whose "prime time" was the morning period, followed in importance by "daytime."

Because the CPB was focused on television in its early days, there was little interest in developing a research apparatus for NPR. However, Stavitsky (1995) argues that a significant change occurred during the 1980s, when audience research was being seen as a way to grow listeners' financial support. This practice was certainly related to the economic pressures facing NPR's managers. NPR was recovering from financial crisis, and the network had agreed to a goal of doubling the audience by 1990 (McCourt 1999). A review of internal documents indicates that there was a significant shift toward thinking of listeners primarily in terms of their economic capacity. By using audience data, NPR and its member stations could identify not only the most likely listeners, but also the most lucrative.

During this period, NPR began to draw upon techniques used widely by the private sector. Specifically, public radio managers began to engage in the practice of market segmentation, in which the aim is to identify and delineate relatively homogeneous segments, which are identified by some common characteristics. These characteristics are useful in explaining and in predicting consumer responses and provide a way in which marketers can effectively channel their resources.

During my interviews, I spoke with the president of an NPR member station in a midsize metropolitan market. He recalled the moment, early in his professional career, when NPR began the pivot toward a greater use of audience research to inform programming: "I think with public radio, with this group of people you would look at and say, 'this is not a very strategic group of people.' But they were actually incredibly strategic. And I think in the late eighties, there started to be this focus on audience. I think it was AUDIENCE 88. And I think they started to realize that they could grow audience. It was a very slow build."

The participant is referring to NPR's AUDIENCE 88 report (Thomas and Clifford 1988), generated by Giovannoni's ARA. The report was designed to help station managers understand the NPR listening audience, which could then inform programming decisions. As the authors of the report stated, it was "an effort to pierce the veil between the broadcaster and the listener, and to capture the clearest possible picture of the people who welcome public radio into their lives." It was the expectation, however, that, by articulating who listened to NPR, station managers could make programming decisions that would yield greater annual funding. According to the participant, this research drove NPR's move toward delivering news over other kinds of programming. As he told me, once NPR moved in this direction, "they saw the audience just swell. And I think it became a recognition that operating as a variety format, kind of the way that public television did, was not going to work."

In a separate interview, I spoke with a station manager who had a long-standing career at various NPR member stations. He also described the moment in his career when NPR began to integrate audience research into their programing. He described this process as a shift from informal decision making to a more formalized process. According to the participant:

> There was a move to adopt the tools of commercial broadcasting, so that we would be more listener centric. I had a sign on my door that said, "Think Audience." Because so much of radio had been internally focused to

the station. . . . So, we started looking at our audience numbers. When I started, it was inconsequential. What we knew was from our best friends and our moms. And so we were looking at ways to increase audience. And the motivation was the threat of killing federal funding. I think that may have factored into it.

This focus on audience measurement now appears to be standard operating practice. "Now it's basic stuff for a program director," the station manager told me. "Knowing that stuff inside and out." Early on, however, NPR had to make a coordinated effort to persuade station managers to rely more on audience research. NPR executives engaged in what Stavitsky (1995) describes as an "audience research road show," in which CPB Research Director Leon Rosenbluth held a series of seminars for station managers on the value and function of audience research. However, the decision to reach a narrowly defined audience was framed as a practical rather than an ideological consideration. NPR researchers argued that it was a simple fact that, because public radio stations existed alongside thousands of commercial competitors, it could only reasonably serve a small portion of America's radio listeners. In the report, ARA expresses the following sentiment:

For much of their history, public radio stations defined their mission in terms that were highly idealistic, broadly inclusive, frequently paternalistic, and often naive with respect to the opportunities and limits of radio broadcasting. Most stations' missions were, at bottom, only vague directives for actual operations, seldom translated into measurable standards suitable for performance evaluation. As audience researcher Tom Church put it, many stations could fulfill their mission without so much as a single person ever listening. (Thomas and Clifford 1988)

In their report, station managers were discouraged from adopting what NPR termed a "diverse-appeal approach," or trying to appeal to too many audiences. NPR claimed that adopting this approach would undercut listener satisfaction and reduce the number of listeners and their level of support. Instead, NPR managers argued for what they called a "unified appeal" (Thomas and Clifford 1988, 34), which they argued would result in a more satisfying and important service for current listeners.

Who, then, would NPR serve? According to NPR's analysis, the listening public could be divided generally into two types of audiences: those who listen to public radio and those who do not. Giovannoni argued that NPR should concentrate its efforts on those who are already likely to listen to NPR. But it was especially important to NPR's economic well-being that stations target listeners who were most inclined to give. ARA's research suggested that listeners were more likely to send money to public radio when they relied upon its service and considered it important in their lives. Furthermore, they found that certain listeners would give if they believed that their support was essential because of limited government and institutional funding.

In an effort to better understand that listener, NPR turned to techniques that were being used in the marketing industry. For example, the network began by using the logic of market segmentation, in which the audience for particular goods can be subdivided by geographic, demographic, psychological, psychographic, or behavioral variables. Defining audiences by factors such as age, gender, and so forth had become a common practice employed by marketers to help define their target audience profiles (Tynan and Drayton 2010).

AUDIENCE 88 affirmed several demographic characteristics of public radio listeners that had been reported in previous studies. First, NPR found that public radio listeners were significantly better educated than the U.S. population as a whole and that this educational attainment correlated highly with income and profession. Over half of public radio's listeners held professional, technical, managerial, and administrative positions. Furthermore, public radio listeners tended to be between thirty-five to forty-four years of age, which NPR claimed was America's best-educated age group (Thomas and Clifford 1988).

But ARA argued that it is not uncommon for people who fall within the same demographic profile to act in radically different ways. Consequently, they went beyond demographic information and began to consider psychographic variables. To better understand who was actively listening to NPR, researchers turned to Values and Lifestyles (VALS), a market research tool for determining the motives behind consumer purchasing decisions. Unlike other segmentation schemes organized by geography, age, or other demographics, VALS accounts for consumers' interests, habits, attitudes, emotions, and preferences. Furthermore, VALS segments consumers into distinct types, or mindsets, based on a specific set of psychological traits and key demographics that drive consumer behavior.

The AUDIENCE 88 survey revealed that NPR appealed to a particular values and lifestyle personality type: Inner-Directed and Societally Conscious (Thomas and Clifford 1998, 4). According to ARA, these listeners have a strong sense of social responsibility, and are inclined to act on their beliefs. They were interested in arts and culture, enjoyed reading and the outdoors, and watched relatively little television. While this audience accounted for only 11 percent of the U.S. population; they were 41 percent of the public radio audience (Thomas and Clifford 1988).

By 1998, NPR's reliance on audience data had become commonsense thinking. Ten years later, when NPR released its AUDIENCE 98 report, a follow-up study to AUDIENCE 88, the network reasserted its claim that NPR is a niche product within the larger broadcasting enterprise, claiming that "each radio station must serve a demographic segment of society—a niche—if it is to compete in the highly fragmented medium. So, by definition, no station's audience can or should 'represent' the entire population" (ARA 1998). However, there appeared to be increased emphasis on mining the listener for economic support. "Ten years ago, AUDIENCE 88 developed the 'programming causes audience' fundamentals of public service—the focus of our activities back then," the authors of the report wrote. "Today, as listener-sensitive income surpasses all other revenues combined, AUDIENCE 98 focuses—necessarily—on the 'programming causes support' fundamentals of public service" (ARA 1998, 1). But station managers were explicitly encouraged to focus on their most lucrative audiences at the exclusion of others. ARA put it bluntly: "listeners who have more money can give more money" (ARA 1998, 4).

While the precise labels shifted somewhat, the essential profile of the NPR listener remained the same. That year, ARA researchers categorized NPR's ideal listeners as "Actualizer-Fulfilleds," those who listen more, give more, and are more likely to have a "sense of community" for public radio than any other listener. The report indicated that seven in ten have advanced degrees, and virtually all have graduated from college. They have a strong sense of civic responsibility, which makes them especially likely to support public radio. Half are current "givers," and of these, two-thirds of this group contributed at least $50 per year. NPR noted that "most important, they can afford it: These middle-aged listeners (average age: 50) have an average annual household income over $100,000" (ARA 1998, 9).

The Actualizer-Fulfilleds were defined in opposition to Strugglers and Believers, those who were listening less and who were more likely to be retired, unemployed, or have no more than a high school education. According to the report, Strugglers are constantly engaged in a fight to make ends meet, and Believers' attitudes and lifestyles make them, in many ways, the opposites of Actualizers. But stations were reassured that this listener was not important anyway, because they were not essential to NPR's bottom line. According to ARA, "this isn't a big deal: these two groups combined comprise less than 10 percent of public radio's cume."

At times, ARA's framing of listeners who do not contribute financially borders on contemptuous. At one point, the authors of the report distinguish between the "Giving Core" and what they call "the Rotten Core" (ARA 1998, 54), an audience that tunes in less frequently, listens between four and four and a half hours less each week, and is significantly less loyal. While both are likely to say that public radio is important in their lives, those in the Rotten Core are much less likely to have a "strong" sense of community with public radio. ARA also argued that this group was less likely to possess the proper beliefs that are associated with giving to public radio.

By the time AUDIENCE 98 was released, it was evident that NPR continued to underserve Black and Latinx listeners. In 1988, NPR found that 91 percent of its audience was white, but only 6 percent was African American, only 2 percent was Asian, and only 1 percent was "Hispanic." That year, U.S. Latinxs accounted for 8.1 percent of the population (*New York Times* 1988). Ten years later, there were only modest gains. In 1998, NPR estimated that 85 percent of their audience was white, but only 8.8 percent of its audience was African American, and only 3.5 percent was "Hispanic." Latinxs accounted for 11.5 percent of the population at the time and were clearly showing a growth trend (U.S. Census 1999). This meant that NPR executives would have to increasingly justify how NPR was serving the needs of Latinx listeners.

The Latinx Listener as an Evolving Construct

When I spoke with NPR's Bill Siemering about how the network originally conceived of its listener, he affirmed that NPR was initially designed to serve ethnically diverse audiences. However, Siemering's conception of diversity was centered primarily on Black and Indigenous communities. This orientation came largely from his own professional experiences. Siemering had previously served as general manager of college radio station WBFO-FM in Buffalo. There, he spent his first years at the station learning about the local community, conducting interviews with the African American community, which were used to develop a series called "To Be Negro." Siemering also worked with Indigenous communities living at nearby Niagara Falls to produce a series of programs on the Iroquois Confederacy called "A Nation Within a Nation" (Janssen and Behrens 2001).

Siemering admitted that Latinxs were not much of a consideration when he wrote his mission statement, stating "at that time, there wasn't much awareness about Latinos." This is to be expected. In 1970, when the network first aired, Latinxs accounted for only 4.5 percent of the total U.S. population (U.S. Census 2012). But when I asked Siemering how a single network was meant to appeal to the broad spectrum of the nation, he stated that a unifying trait of NPR's audience is curiosity. "Being curious is very important," Siemering told me. "That cuts across all divides."

The notion of curiosity has been a defining characteristic of the NPR audience over the course of its history and is reflected in the marketing materials NPR uses to sell its audiences to corporate underwriters. For example, NPR markets a number of products under its "Curious Listener" series, which educates listeners on how to appreciate music and culture. However, Siemering was firm in his belief that NPR should not consider the economic value of its listeners to be its paramount consideration. When we spoke, he read aloud a sentence in the mission statement that he felt was particularly important: "NPR would not regard its audience as a market." Yet, this is exactly how the network began to regard the listener. The research conducted by ARA researchers was designed to cultivate a listening audience that would support the network financially. This strategy has, in turn, informed how NPR conceives of, and pursues, its ideal Latinx listener.

There is little specific mention of Latinxs or Hispanics in *Public Television: A Program for Action*, the Carnegie Commission's (1967) report on public broadcasting, which prompted the passage of the Public Broadcasting Act in 1967. This was likely due to the fact that the invention of a unified pan-construct termed Hispanic or Latino had not yet been widely employed by marketers, politicians, and activists (Mora 2014). However, the report makes clear that public broadcasting is meant to serve diverse audiences: "America is geographically diverse, ethnically diverse, widely diverse in its interests. American society is proud to be open and pluralistic, repeatedly enriched by the tides of immigration and the flow of social thought. Our varying regions, our varying religious and national and racial groups, our varying needs and social and intellectual interests are the fabric of the American tradition" (Carnegie Commission 1967, 14).

Similarly, there is little direct mention of Latinxs in *The Hidden Medium* report (1967), which made the specific case for public radio. However, the report did make reference to disparate Spanish-speaking communities across the country, which could be served by particular public radio stations. These communities were collapsed into the larger category of "minority groups," referred to in other parts of the report as "special audiences" and "forgotten audiences." Often framed as society's outliers, these groups were characterized as problems that educational radio could solve. For example, the authors argued that there are plans for public radio to serve "the disadvantaged," which included "the elderly, the chronically ill, the poor, the migrants, the retarded, the ethnic and racial minorities" (Herman W. Land Associates 1967, VIII-2).

The authors also wrote that public radio could provide news for non-English-speaking listeners, referring to a pilot program that had been proposed to serve what they termed "the Spanish-speaking illiterates in Florida" (Herman W. Land Associates 1967, VIII-4). Based out of an elementary school in Delray Beach, the station was intended to be operated by migrant workers in southern Florida, who could "make their needs known and understood." Programming would be developed and produced by the community and broadcast both in

Spanish and English, offering lessons on conversational English as well as tips on cooking, baby care, health care information, and farm labor news.

Almost a decade after NPR was established, the Carnegie Commission created a second blue-ribbon panel to assess the development of the U.S. public broadcasting system. In its report titled *A Public Trust: The Landmark Report on the Commission on the Future of Public Broadcasting* (1979), the commission affirms the notion that public media's economic interests should not supersede its social mission. This meant that NPR should reach a diverse audience, particularly those ignored by commercial media:

> Unlike commercial radio and television, most print media, and many new communications services, public broadcasting creates programs primarily to serve the needs of audiences, not to sell products or meet the demands of the marketplace. This ideal demands that public television and radio attract viewers and listeners whose tastes and interests are significant, but neglected or overlooked by media requiring mass audiences. The noncommercial nature of public broadcasting has important implications for its programs, its relations with creative talent, and its mission to underserved audiences. (Carnegie Commission on the Future of Public Broadcasting 1979, 25)

As early as 1977, the Carnegie Commission realized that NPR had seriously fallen short of its diversity mandate, stating: "the effect of this history is a public radio system that does not reflect the pluralism that is a highly valued characteristic of American society" (Carnegie Commission on the Future of Public Broadcasting 1979, 193). In January of 1977, the CPB commissioned a task force to examine how public media was faring in addressing the needs of minority publics. Led by Professor Gloria L. Anderson, vice chairman of CPB, the twenty-eight-member coalition met to develop policies that would maximize the growth, development, employment, and participation of minorities in all aspects of public broadcasting. In the words of the committee, their goal was to "set out to determine how well that system by then known as public broadcasting has met, and is meeting, the informational and educational needs and interests of Blacks, Asians, Latinos and Native American citizens" (Center for Public Broadcasting 1978, xiii).

After eighteen months of deliberations, the committee issued a scathing report titled *A Formula for Change: The Report of the Task Force on Minorities in Public Broadcasting* (CPB 1978). In it, the committee identified several barriers to the full participation of minorities in public broadcasting and identified ways by which to overcome them. In its executive summary, the task force described the public broadcasting system as being "asleep at the transmitter" (Corporation for Public Broadcasting 1978, xiii), and the committee found that minority programming was seriously deficient and that proposals for minority programs were rejected at a disproportionately high rate.

In their report, NPR's track record with respect to minority programming was described as "appalling" (Corporation for Public Broadcasting 1978, xiii). The task force found that, in 1975, NPR distributed only sixty-one hours of original minority program hours, the production costs of which were approximately $65,098. This accounted for only 4 percent of the total NPR programming budget for that year. The network had made only minor progress in two years. By 1977, only seventy hours of programming distributed by NPR were devoted to programs by or about racial and ethnic minorities. The committee found that the scarcity of minority programs was directly related to the insufficient number of minorities who were employed in public broadcasting, with a particular lack of minorities in decision-making positions. Over half of public radio licensees had no minority employees.

To address these deficiencies, the committee laid forth a number of recommendations, which focused on increasing minority ownership, reforming the network's hiring practices, and investing in program development. But the commission also urged NPR to review its research practices so that it could better measure minority audiences, arguing that nonwhites are underrepresented in the national samples used in syndicated research. Since current research methods had proven to be inadequate, the committee recommended that funds be allocated by CPB, NPR, and PBS specifically to develop specialized research on minority audiences. Furthermore, the committee recommended that this research be conducted by professionally trained minority personnel.

According to Berkman (1980), the CPB was unresponsive to the task force's recommendations, prompting its members to disband it. For example, despite the task force's call for a commitment to research, both NPR's AUDIENCE 88 and AUDIENCE 98 reports reveal little insight about Latinx listeners other than their designation as "Hispanics." This may be attributed to the lack of industry research on Latinxs overall. The first industry study of the Hispanic market wasn't published until 1981. Titled "Spanish USA: A Study of the Hispanic Market in the United States," the studies were conducted by Yankelovich, Skelly, and White, Inc. on behalf of the Spanish International Network, which would later become Univision, the largest Spanish-language network in the United States.

The Spanish USA study exemplifies the way in which audience research has been used to meet industrial needs. Sponsored by a commercial Spanish-language network, the research was intended to sell the idea of a Hispanic market to prospective advertisers. To accomplish this, the report does the ideological work of reconfiguring different communities with distinct linguistic and cultural characteristics into a unified, pan-ethnic market, thereby minimizing differences between Puerto Ricans, Cubans, Mexicans, and other groups (Chávez, 2015). The Yankelovich, Skelly, and White study was also significant because it legitimized the importance of the Spanish language in unifying the Latino community.

But, while the Spanish-language broadcast industry attempted to frame the Hispanic market as a unified, Spanish-speaking bloc, NPR was moving in the opposite direction: separating the Latinx community into distinct segment types so that it could target the audience of Latinxs who were most congruent with their current listeners. Based on their research for the AUDIENCE 88 and 98 studies, NPR found that Blacks, Latinxs, and other ethnic groups were underrepresented in NPR's preferred VALS types and that it would require significant resources to achieve "proportionate" service to Blacks, Hispanics, and other ethnic groups (Thomas and Clifford 1988, 35).

Rather than investing in original programming that might bring in minority listeners, as the CPB task force recommended, NPR now argued that it could grow its Latinx audience simply by staying the course. In a 1989 CPB research summary, ARA's Frank Tavares made the case that NPR could succeed in courting Black and Latinx listeners with its current programming strategy. Drawing from a variety of sources, including Arbitron, a custom survey used for the AUDIENCE 88 study, and an extensive review of the literature on minorities and radio, Tavares set out to challenge three presumptions: that public radio does not appeal to minority listeners; that Latinxs are a single, unified group with shared characteristics; and that Spanish is a unifying trait.

Tavares acknowledged that the minority population of public radio was lower than that of the general population. NPR data showed that Black and Latinx listeners were less likely to listen to public radio, and when they did, they did not listen as long. However, Tavares argued that NPR's current numbers did not necessarily reflect a categorical rejection of public radio, but simply that the right kinds of Latinxs were not exposed to the right kind of programming. According to Tavares (1989):

> The average Black and Hispanic listener chooses radio
> stations for the same reasons other radio listeners do. The
> stations meet particular needs. They meet those needs
> by broadcasting in certain formats. Listeners whose in-
> terests lay outside of the format of a station are unlikely
> to use that station. This has great implications for those
> of us in public broadcasting wanting to reach Black and
> Hispanic listeners.

In an implicit critique of Spanish-language media, Tavares argued that stations should consider the diversity of the Latinx population, stating that "the tendency of non-Hispanics to lump all Latino cultures together is well known" (Tavares 1989, 5). According to Tavares, Latinxs could not be understood as a unified group, and their preferences for radio formats varied tremendously from market to market. Similarly, he cautioned station managers against equating *Hispanic* with *Spanish-speaking*, arguing that while Spanish-language formats may be predominant in one media market, they may be less prevalent in others.

Because of the diversity of the market, Tavares called for a conservative programming strategy, since, as he put it, "the (minority) listeners we want to reach are already in the mainstream" (Tavares 1989, 9). Rather than create specialized programming that might fragment the listening audience, station managers should recommit to their current formats. Because NPR member stations were almost exclusively monolingual spaces, this meant excluding Spanish-language programming.

Tavares' recommendations reflect a programming strategy that NPR had already committed to in its AUDIENCE 88 report, which was to avoid Black- and Latinx-oriented programming. By the time the network issued its AUDIENCE 98 report, NPR's conservative programming strategy was all but cemented. The authors of the report characterized its programming strategy as a tension between two competing strategies: a "strategy of targeting" and "a strategy of transcendence." ARA argued that when individual stations attempt to serve vastly different audiences, they do not and cannot serve the public. But, curiously, ARA wrapped their argument in the rhetoric of inclusivity, arguing that "the strategy to transcend racial heritage and the strategy to target it are at direct philosophical odds. The targeting strategy emphasizes differences in our racial and cultural backgrounds. The transcendence strategy emphasizes similarities in our characters" (ARA 1998, 30).

According to ARA, the strategy of targeting people by their specific racial or ethnic backgrounds was shortsighted. Not surprisingly, NPR stations were encouraged to adopt a "strategy to transcend," which, according to the network, would benefit NPR in the long-term. "We would be right to accept that most public radio listening is programming that seeks to transcend. This is as true for minority listeners as it is for others," the authors of the report urged station managers. "We would be wrong to compromise this programming's appeal by bending it toward the strategy to target. These two strategies are incompatible in the same program stream. We would be right to assume that the strategy to transcend is well aligned with powerful demographic trends among America's minority populations" (ARA 1998, 30).

The network remained steadfast in its belief that demographic trends favored NPR. "Public radio's minority audience will continue to grow because the college-educated minority population will continue to grow" (ARA 1998, 29) is how ARA's researchers put it. Rather than evolve its programming to have wider appeal, NPR advocated that stations focus on what they already did best: developing programs for highly educated listeners. According to ARA:

> Driven by the strategy to transcend, the single most de-
> fining characteristic of public radio's audience today is its
> college education. If it remains so into the future, public
> radio can expect to serve even more minority listeners
> tomorrow. As the number of well-educated minority cit-
> izens grows, so grows public radio's minority audience.
> (Thomas and Clifford 1988, 32)

When measured by their formal education and their character, researchers asserted that that public radio's minority listeners have more in common with other public radio listeners than with non-listeners who share their ethnic or racial backgrounds. ARA argued that this was significant because, in ARA's words, "it reminds us that listeners are best understood by what draws them to us—and not by where they come from." According to ARA:

> [This strategy] ignores demographic distinctions of age
> and sex, race and ethnicity. In its best moments it tran-
> scends racial and ethnic differences. It focuses on vir-
> tual communities of listeners who share the values and
> attitudes formed by their educational experience. The
> second strategy targets listeners who share certain racial
> or ethnic characteristics. It focuses primarily on persons
> with these characteristics. (ARA 1998, 29)

Despite its pretense of inclusion, NPR's "Strategy of Transcendence" reflects a colorblind ideology, in which NPR would acknowledge race, but would disregard its importance when it came to programming decisions. But as Garcia (2010) has argued, colorblindness has been used as a strategy that has enabled powerful institutions to maintain the dominant social order. In a curious twist of logic, NPR executives then argued that this strategy of focusing on the country's elite, rather than its most disenfranchised, was more in line with its original mission:

Is this what we want? Well, it's what we set out to do thirty years ago—to provide a beacon of public service that places character over color. The character of this beacon is transcendent. It transcends geography with a "sense of community" engendered across vast physical distance. It transcends age and sex. And by operating in the enlightened dimension of education's values and attitudes, it transcends color through its very indifference to it. (ARA 1998, 33)

Later in the report, ARA appropriates the language of the civil rights movement, arguing that NPR programming should focus on the character of its listeners rather than on their ethnic differences. By attending to the needs of minority listeners through dedicated programming, NPR argued that the network would actually be undermining its mission of inclusivity, which was built into its mission statement:

Thirty years ago, public radio set forth a beacon of public service to advance understanding among people of good will; to unite rather than divide; to include rather than exclude; to transcend races and creeds, origins and situations. The mission embodied in this strategy holds as self-evident that a person's character, values, and attitudes are more relevant than one's racial or ethnic background. It emphasizes the similarities among people rather than their differences. (ARA 1998, 29)

There is a clear contradiction between NPR's rhetoric of inclusivity and its actual targeting practices. The network had consistently made the case throughout its internal reports that it should serve, almost exclusively, listeners who possessed significant amounts of social, cultural, and economic capital. Furthermore, the network aggressively pursued a conservative, rather than transformative, approach to programming. Instead of changing their programming formats, NPR member stations were directed to simply seek out the audiences that were already consumers of the existing product.

The disparity between NPR's rhetoric and its professional practices became evident when NPR president Douglas Bennet testified before the U.S. Congress in 1988 to seek increased funding for the CPB, NPR, and PBS. During his opening remarks, Bennet argued that, with sufficient federal resources, public broadcasting could add new programs to serve the needs of NPR's diverse audience. Bennet went on to claim that NPR was engaging in practices that ensured that the network's programing reflected the breadth of our society, testifying that NPR was actively involved in the "production and acquisition of programming by, for and about ethnic and social minorities—Black Americans, Native Americans, Pacific and Asian Americans, women, children, the elderly and disabled persons" (U.S. Congress 1988, 95).

Throughout the deliberations, it is evident that Latinxs had emerged as a constituency of particular interest to politicians. During the testimony, lawmakers pointedly asked about the ways in which CPB, NPR, and PBS "served the interest of Hispanics" (U.S. Congress 1988, 68). In the appendix to the hearing, Arnold Torres, a vocal critic of NPR, had submitted a list of recommendations to Congress, which he believed would ensure greater Latinx representation on NPR. This included the creation of a Hispanic Advisory Committee to work with the CBP and NPR to develop Spanish-language programming. Torres also suggested directing the CPB to develop a comprehensive plan for assessing the TV and radio needs of Hispanics and to report on their progress in meeting these needs.

In his formal response to the senate, Bennet rejected the idea of creating programming for Latinxs, arguing that NPR already reflected many different viewpoints. In direct contrast to his opening statement, Bennet argued that it was not appropriate for NPR to develop programming designed exclusively for any individual group. As Bennet stated: "Programming specifically for Hispanic audiences is no more appropriate than programming aimed at any other specific group. Furthermore, research indicates that viewers and listeners do not use television or radio in any way that is related to their ethnic or racial characteristics" (U.S. Congress 1988, 38).

Bennet's claim that U.S. Latinxs do not use media based on their "ethnic or racial characteristics" ignores the profound demographic and industry changes that were occurring at the time. As Wilkinson (2016) argues, the 1980s were the "Hispanic decade," in which an increase in Latinx immigration led to the growth of the Latinx market. Latinxs had a particularly strong presence in California, Texas, New

York, and Florida. To meet the needs of this growing community, a robust Spanish-language broadcast industry was emerging, led by Univision and Telemundo.

However, NPR continued to press the argument that it should not change its programming strategy in response to the country's demographic changes. Consider the following exchange between Bennet and Republican senator Pete Wilson of California. Wilson would later become infamous as the governor of California who championed Proposition 187, legislation that sought to deny undocumented immigrants health care and education services. During his line of questioning, Wilson inquired about how NPR serves Latinxs, whose presence was significant in California: "On the one hand, I think you probably consider that you have an obligation to respond, to provide programming, that will be of greatest benefit to the greatest number of people. On the other hand, I represent a state that increasingly is attracting new residents, some of them from foreign lands" (U.S. Congress 1988, 147).

Wilson then directly raised the issue of NPR's cancelation of *Enfoque Nacional*. But Wilson's exchange with Bennet seemed less like an interrogation of NPR's practices and more like an opportunity for Bennet to justify the decision. Wilson continued: "I thought I understood you to say that while you thought the show had excellent content, you found by some kind of a ratings service or measurement that of audience reaction, that it was not generating much listenership" (U.S. Congress 1988, 148).

Wilson's setup allowed Bennet the opportunity to restate his case that audience data confirmed that Spanish-language content has no place on NPR. According to Bennet, *Enfoque Nacional* "was in Spanish, and the reality is that almost all the public radio stations in this country are English language stations. . . . So what is the solution to that? What do we think the solution is to format a program so that it will work for radio and then go to where the Spanish language listeners are, the Spanish-speaking listeners, which happens to be on commercial radio in this country."

Bennet's testimony demonstrates the way in which audience research can serve ideological purposes. Presented as verifiable fact, industry research enables media practitioners to make claims about particular communities, their beliefs, and their media practices. These forms of knowledge may be based on the pretense of objectivity, but they are only semi-intuitive in nature (Hackley 2002). Here, NPR employed audience research as a way to give itself permission to pursue an educated, affluent listener, one who would serve its financial interests. Meanwhile, disenfranchised listeners within the United States were relegated to the Spanish-language, commercial realm. However, this strategy would prove unsustainable, as Latinxs would soon grow into a social, economic, and cultural force.

The Latinx Listener as Economic Imperative

In the years since Bennet's testimony, two competing trends have continued. First, the number of U.S. Latinxs has continued to grow. Second, NPR has continued to pursue an audience that is primarily white, affluent, and educated. This has led to an increasing disparity between NPR's audience and the public that it purports to represent.

If anything, NPR's research apparatus has become more effective at capturing its ideal audience and selling that audience to corporate underwriters. Overall, public radio networks have increasingly relied on commercial approaches to audience research, and have invested in digital platforms, which has allowed them to reach niche audiences with more personalized content. Consider how the NPR listener is described by NPM, the unit that solicits corporate sponsorships on behalf of NPR. At first glance, NPR is framed as a way to reach a diverse public. As NPM puts it, "across platforms, public media engages larger, more diverse audiences than ever before—audiences looking for stories of today and a vision of tomorrow."

However, NPM also makes clear that the listener to whom NPR delivers is cultured, engaged, influential, and conscientious. According to NPM, the typical NPR listener enjoys traveling and learning about new places and cultures and has the resources to do so. NPM claims that their listeners are 110 percent more likely than other listeners to work in top management and are 160 percent more likely to be "C-suite." They are 152 percent more likely to have visited an art gallery or show in the past year. They are also civically engaged. According to NPM, the NPR listener is 210 percent more likely than other listeners to have contacted a politician at the state, local, or national level (National Public Media 2020).

At the same time that NPR touts the cultural and economic capital of its current listeners, the network has also expressed a desire to reach Latinx listeners. This has been framed as both a social and an economic mandate. In 2014, the CPB's chief content officer, Joseph Tovares, publicly claimed that NPR would adopt a "diversity as a business imperative" model that would be built into its strategic plan. By attending to the three D's (digital, diversity, and dialogue), NPR would expand its operations into digital broadcasting, diversify and engage its audiences, and grow its net revenue.

However, NPR and its member stations have struggled to reconcile these competing mandates: to continue to pursue a privileged audience, while also reaching out to Latinx listeners. I found that station managers were most attentive to the immediate goal of maintaining their economic viability by securing their current listener base. By contrast, expanding their listener base by targeting Latinxs seemed like a secondary, longer-term strategy. For example, I spoke with the chief content officer at one NPR member station in a midsize metropolitan market. I asked him about the audience for whom his station was creating programming:

> The median age there is probably early to mid-fifties . . . it's younger than you might expect, I guess. But I will say that one very strong characteristic is education. If you have a college degree, you're far more likely to listen to public radio than other formats. And affluence. Generally, the audience skews towards having more income. Although certainly we're not programming to get that audience, that's just the audience that gets here.

When I asked the participant how the station fares with minority audiences, he indicated that the station did not perform well. But he believed that was a function of the region's demographic, rather than the content they produced or distributed: "We have some data on the ethnic breakdown of the audience, but it's tough in this area because the population is overwhelmingly Caucasian. But we at least have a measurable amount of a Hispanic audience and almost measurable amount of an Asian American audience."

This is consistent with other interviews I conducted at the station level, which suggested that engaging Latinx listeners is desirable, but not critical. Furthermore, I found that efforts to diversify must be taken without isolating NPR's core audience of older, wealthy listeners, what one news director referred to as their station's "legacy audience." According to the participant, "we have an older cohort who is going to continue to consume media the way they always did. And we've got to continue to be there for them . . . we have to be very cognizant of not destabilizing our older donor base." The participant was candid about the economic concerns that drive his station's programming. In an effort to maintain consistency, stations are encouraged to appeal to Latinx listeners in ways that require little change to the programming schedule. But I also found that efforts to integrate Latinxs, however modest, can be met with resistance both internally and by members of the listening public.

Southern California Public Radio: an LA Story

I had a chance to visit NPR member station 89.3 KPCC, Southern California Public Radio (SCPR), which serves the greater Los Angeles area and is located in Pasadena, California. KPCC had recently gained notoriety as an exemplar in the industry of how to integrate Latinx listeners. In 2014, the Latino Public Radio Consortium (LPRC) worked with KPCC to document the process.

In 2000, Pasadena City College, the station's original licensee, found that it was costing more money than the college wanted to spend. In an effort to offset costs, the college issued a request for proposals (RFP) for outside operators to manage the station, eventually turning their operations over to SCPR, a subsidiary of American Public Network. SCPR recognized that, to succeed in Los Angeles, with a significant Latinx population, it had to make it a strategic priority to engage Latinx listeners (Latino Public Radio Consortium 2014).

To accomplish this goal, the LPRC found that the station made significant change, focusing on community engagement, investing in changes to its programming, and reshaping its organizational culture. Strategically, SCPR's leadership board actively recruited members with diverse networks that could be activated in support of the station. To accomplish this, the station hosted a number of community events and held focus groups and meetings with leaders in the Latinx community.

KPCC also experimented with adopting best practices from commercial radio, which included a greater use of humor and a more energetic pace, distinguishing KPCC from the slower, more deliberate pace of most NPR member stations. To achieve this new sound, KPCC invested in hiring journalists, producers, and hosts who were Latinx and/or came from commercial backgrounds. Rather than suppress their ethnic identities, there was the expectation that KPCC staff would report on stories they had personal insights on. According to the LPRC report, KPCC was seeking talent that "had both the linguistic and cultural ability to connect with the communities they report on" (Latino Public Radio Consortium 2014, 10). As part of their hiring strategy, KPCC was also looking for talent that was technologically savvy and was comfortable promoting stories on social media.

According to the LPRC report, the station's managers were cognizant that these changes would cause disruption both within the organization and amongst its listeners. Anticipating this disruption, KPCC recognized that they needed to change the organizational culture and clearly communicate their vision to all their employees. Expecting resistance to these changes, they invested in numerous training and team-building exercises that would help staff thrive in the new environment.

During my conversation with KPCC president Bill Davis, we discussed the unique challenges of creating programming in a city like Los Angeles, which is 49 percent Latinx. Davis acknowledged that framing Latinx issues as minority issues in a market like Los Angeles is misguided, stating "everything in Los Angeles is a Latino issue." But Davis was also keenly aware that, while his station serves an important civic function, it is also a commodity in the media marketplace.

According to Davis, the "Spanish-speaking audience is overserved" in Los Angeles, given the number of Spanish-language radio stations competing in the market. In an effort to find its niche amongst listeners, therefore, the station had to narrowly define the Latinx audience they wanted to pursue. In doing so, the station engaged the logic of market segmentation, or the practice of dividing a given population into distinct subgroups, which are presupposed to have similar attitudes, behaviors, and lifestyle characteristics. This practice allows the organization to efficiently channel resources to the group that will be most responsive to their efforts.

According to the LPRC report, both the CPB and KPCC invested heavily in research that would help them determine the specific segment of the Latinx community on which to focus. The research broke up the Los Angeles Latinx population into three segments: recently arrived immigrants, first-generation Latinxs, and second-/third-generation Latinxs (Stuart 2012a). Immigrant Latinxs were found to have little interest in news and felt well served by commercial Spanish-language radio. By contrast, first-generation Latinxs were active consumers of news and music, but had almost no awareness of or interest in public radio. Focus groups helped determine that late second-generation and early third-generation Latinxs are most likely to prefer English-language news and information services, rather than Spanish or bilingual programming.

Ultimately, it was decided that the station would best be served by creating programming for second- and third-generation Latinxs, which, the research indicated, were English-dominant and already civically engaged. As Bruce Theriault, then vice-president of radio at the CPB, stated at the time, this listener is "well informed and has a strong desire for news programming that presents multiple perspectives of an issue" (as cited in Stuart 2012a). The research further indicated that this audience was amongst the most frustrated media consumer in Los Angeles. Their frustration was twofold: (1) they found that Spanish language media lacked useful news and information, and (2) they felt ignored by English language media.

By some measures, this strategy has been a success. KPCC claims that it was able to integrate Latinx listeners without isolating its legacy audience. According to the LPRC report, listener support nearly doubled from $6.5 million to $11.4 million, and corporate underwriting revenue increased from roughly $5.3 million to $7.8 million between 2009 and 2014. However, my interviews reveal a more complicated picture. The integration of Latinxs was met with a tremendous amount of resistance from listeners, journalists, and the station's employees, which played out publicly in the local media.

Consider the case of A. Martínez, host of *Take Two*, a morning news and culture show featured on KPCC. In 2012, Martínez was originally recruited to serve as co-host for the popular *Madeleine Brand Show*, which was later renamed *Brand and Martínez*. According to Martínez, his hiring was met with some resistance from some listeners and the local media, some of which had racial undertones. "Do you know what swarthy means?" Martínez asked me during our interview in Pasadena, California. "Because I didn't. I had to look it up. It means dark skinned." Martínez was referring to an article written in *LA Weekly*, which stated that Madeline Brand would be "bantering with a swarthy new co-host" (Stuart 2012b). According to Martínez, the scrutiny he received upon his arrival at NPR caught him by surprise: "I didn't realize this would make news. Someone gets hired at a radio station, why is that news? It turned out to be huge news because of the directive of the grant, what they were looking for. [People believed] 'well we found a brown dude, let's just stuff him in there.' That made news."

Martínez is referring to the One Nation Media Project, which provided the grant that made his hire possible. The goal of the organization is to provide coverage to multiethnic communities in Southern California. KPCC had applied for the grant to diversify its newsroom, but Martínez suspects some members of the community may have perceived this as, in his words, "forced affirmative action." According to Martínez, his legitimacy as an NPR host was immediately suspect: "My education was questioned. 'He went to three junior colleges. He went to a state school. He's not a Berkeley grad. He's not a J school grad from Columbia, or from NYU or from Northwestern. Does he have the intelligence to do this? Is he brown enough? Is he not brown enough?'"

Unlike his counterpart, Brand, who was professionally groomed in public radio, Martínez began his career in sports radio, which was seen as incongruent with the NPR style. Ultimately, the collaboration proved to be unsuccessful, and, after four weeks, Madeleine Brand left KPCC, citing creative differences. Fans of the show, however, were critical of the station's Latinx outreach efforts. In 2012, *LA Weekly* ran an article titled "How KPCC's Quest for Latino Listeners Doomed the Madeleine Brand Show," in which author Tessa Stuart laments the loss of the show. "Gone was Brand's theme song by indie band Fool's Gold," Stuart wrote in her essay. "In its place was the trilling pan flute of 'Oye Mi Amore' by Maná, a Mexican rock group whose popularity peaked in the 1990s. Suddenly, instead of the usual segments about *Downton Abbey* and disputes at a Brooklyn co-op, there was a segment about the death of a tortilla magnate, followed by one on Hatch chile season" (Stuart 2012a).

With her references to *Downton Abbey*—the British period drama—and indie bands, the reformatted show was clearly an affront to Stuart's elite taste sensibilities. But in her assessment of KPCC, Stuart framed the station's pursuit of Latinx listeners as economic opportunism rather than civic responsibility. According to Stuart, the station's diversification strategies were driven by the desire to increase ratings and pursue grant dollars, which she believes is misguided. The product of such a strategy, she argues, will be programming that appeals neither to Latinxs nor to NPR's current audience.

In her critique of KPCC, Stuart articulates the challenge faced by many station managers operating in markets where there are significant Latinx populations, which is how to integrate Latinx listeners without isolating their core audience of listeners. In Los Angeles, KPCC, with resources from the CPB, made a concerted effort to invest significant resources to modify its programming, but these yielded only modest change. KPCC programming remains almost exclusively in English, despite the prevalence of Spanish in the area. Furthermore, we must recognize that the strategic framework from which KPCC was operating ensured that the station would not serve the area's most disenfranchised Latinx listeners. By focusing on Los Angeles's most engaged Latinxs, KPCC in essence has accomplished what Frank Tavares had advocated for almost thirty years earlier, which was to make modest changes to format and to pursue a Latinx listener who could fit relatively comfortably within their current audience profile.

WHOSE IS THE VOICE OF THE AMERICAN PUBLIC?

The man who realizes that the pronunciation of the loudspeaker is not his own, and not one that he hears about him in his everyday life, may resent the fact that an alien dialect is inflicted upon him.

—A LLOYD JAMES,* BBC 1926.

On April 24, 2019, NPR public editor Elizabeth Jensen (2019c) wrote an opinion piece titled "You Say Bogota, I say Bogotá," in which she defended NPR's policy regarding on-air pronunciation of non-English words. Her essay was written in response to feedback raised by listeners, who had expressed concern that NPR was engaging in a form of language-based discrimination by choosing to pronounce some formal names and titles in Spanish. While NPR will occasionally pronounce words as spoken in their country of origin, Jensen noted that, of the complaints she received, the listeners expressed concerns only about accurate Spanish pronunciations (Jensen 2019c).

In the essay, Jensen made the argument that NPR, under limited conditions, should reflect the diversity of the nation while respecting the linguistic practices of its hosts and correspondents. In doing so, she referenced NPR correspondent Lourdes Garcia-Navarro, who, earlier that month, had taken to Twitter to defend her choice to code-switch, choosing to pronounce some words in Spanish while leaving other words anglicized. In her post, Garcia-Navarro defended her position, stating "I'm bilingual. I'm not affecting an accent or trying to be pretentious. I speak the language. I think people aren't used to hearing things pronounced correctly, because we are not used to hosts from different backgrounds on the air" (Navarro 2019).

Jensen's position that NPR should necessarily challenge listener's linguistic sensibilities seemed to be an evolution of NPR's previous

---

*From Lloyd James's essay "Broadcast English," The English Language, Volume 2: Essays by Linguists and Men of Letters, 1858–1964. Cambridge University Press.

position. In a 2005 essay titled "Pronunciamentos: Saying it Right," Jensen's predecessor, Jeffrey Dvorkin, appeared to make a different argument, which was that foreign words must necessarily be anglicized in order to facilitate clear communications. In it, Dvorkin posed the question: "Should we say 'Pah-REE' instead of Paris? The former is linguistically correct, but that sounds *très* pretentious to American ears. And perhaps this is an important point not to be lost: NPR is in the business of communicating news and information to a primarily American audience" (Dvorkin 2005a).

While Jensen's essay is meant to signal progress, the very premise of her piece reveals vastly different assumptions about the nature of public radio. First, the essay affirms the belief that NPR has an identifiable on-air speaking style, one that is generally devoid of regional and ethnic accent, what sociolinguists refer to as Standard American English (SAE). Second, the perception that Spanish pronunciations would be seen as problematic suggests that there is a perceived incompatibility between NPR's on-air speaking style and everyday Latinx speech.

It is important to recognize that, while certain formal names may be tolerated on NPR, the network remains a highly standardized space in which editorial decisions are made in favor of a presumed listener who is an English monolingual. In this space, Garcia-Navarro may be permitted to pronounce formal names in Spanish, but she does not have the capability to report in Spanish, a language in which she is fluent. In Jensen's essay, Mark Memmott, NPR's standards and practices editor, framed this as a choice between linguistic diversity and clear communication. "We want our correspondents and hosts to speak naturally," Memmott stated, later adding, "as long as it's understandable to the bulk of the audience."

The lack of linguistic flexibility that is afforded to Latinx public radio practitioners connects to the larger issue of how Latinx speech is policed in white public space. As Urciuoli (1996) argues, linguistic mixing may be tolerated, and even appreciated, among Latinxs, but it is highly restricted in spaces where Latinxs are at a structural disadvantage. Scholars, primarily from the field of sociolinguistics, have argued that broadcast institutions play an important role in promoting linguistic homogeneity, but NPR raises unique considerations. As the nation's largest publicly funded radio network, NPR has the unique mandate to better represent the linguistic diversity of the nation.

In this chapter, I explore the degree to which NPR's linguistic practices shape the way in which the network serves the broader spectrum of the American public. Specifically, I employ language ideology as a lens through which to examine the processes that maintain a standard language within public radio and how such practices inadvertently shape civic discourses. Here, I argue that, despite its pretenses of social inclusion, NPR has defined the voice of the American public in highly limited ways, a process that is maintained by constant coercion. I also argue that, by constraining linguistic diversity, NPR further marginalizes those most in need of civic information.

The Voice of the People

In an essay titled "Broadcast English" (1926), Arthur Lloyd James, a lecturer in phonetics at the University of London, addressed two conflicting realities regarding the use of standard English in public radio. Written on behalf of the newly formed BBC, James acknowledged that variation and linguistic diversity are intrinsic to all spoken languages. Different ways of speaking are governed by local conventions and are grounded in their own historical contexts. Furthermore, one's way of speaking reflects a number of social identities to which one belongs. At the same time, a national broadcast system creates the practical need to use a common language that can connect all parts of a nation. Written at the onset of radio, James described this paradox:

> We now have a certain type, or rather a carefully chosen
> band of types of English, broadcast over the length and
> breadth of our country, so that although many listeners
> hear daily a type of speech with which they are familiar,
> and which they habitually use, many others hear a type
> that is different from that which they usually hear and use.
> This is in itself is enough to ensure abundant criticism.

James recognized that choosing a broadcast standard would privilege some speakers while isolating others. But this raises a question—whose way of speaking will become representative of the nation? Here, the BBC was unflinching. Seen primarily as an educational resource that could elevate the masses, the BBC promoted British Received

Pronunciation (RP) as its broadcast standard, an accent used by a small percentage of the British population. By specifically focusing on a spoken variety used by London's elite, the BBC intended to use broadcasting as a social model by instructing its listeners in "proper" English (Schwyter 2016).

To ensure that the BBC acted in this capacity, the corporation formed the Advisory Committee on Spoken English, whose members included British luminaries such as poet laureate Robert Bridges, who served as chairman, with playwright George Bernard Shaw serving as vice chairman and James serving as secretary (Cassidy 1992). However, Mugglestone (2008) argues that the committee's work on behalf of the BBC was filtered through discourses of prejudice, in which nonlocalized standard English was equated with civilization and power—the King's English. By contrast, regional linguistic styles were associated with cognitive deficiency or characterized as a corruption of true English.

Schwyter's (2016) research on the BBC's Advisory Committee on Spoken English reveals that, as early as 1934, "foreign words" were considered to be a "source of anxiety" (Schwyter 2016, 37), and it was the committee's policy to anglicize as many foreign words as possible. For example, Schwyter describes the committee's discussion around the word *ski*, a loanword that was borrowed into English from Norwegian. While the pronunciation *skē* had become widely used in Britain, *shē* was also accepted usage at the time. However, the committee had taken the position that multiple broadcast pronunciations on BBC would be confusing to listeners. Furthermore, the committee was confident that *ski* (*skē*) had been used in Britain long enough to be "given papers of naturalization" (Schwyter 2016, 51).

When the United States Congress passed the Public Radio Act of 1967, the nation's first public radio network was designed to reflect more populist sensibilities and, therefore, carried different assumptions about language. Unlike the BBC, which was developed primarily as an education and entertainment resource (Schwyter 2016), NPR was intended to serve a democratic function, tasked with engaging audiences in civic discourses. This called for NPR to rely more heavily on the American vernacular and to reflect a more diverse representation of the American electorate that went beyond the educated elite.

In NPR's original mission statement, Bill Siemering was optimistic about the unique possibility for public radio to represent the nation's diversity. As a primarily auditory medium, radio has the potential to capture the range of the human voice: "It [NPR] would not, however, substitute superficial blandness for genuine diversity of regions, values, and cultural and ethnic minorities which comprise American society; it would speak with many voices and many dialects."

Despite this promise of linguistic diversity, NPR's architects had chosen as its standard a way of speaking that is not overtly associated any particular social group, but more broadly with the leveled dialects of the Northern Midwest. It is a dialect where regionally and ethnically marked features have been suppressed so that they are commonly perceived as neutral. This notion of a relatively universal and socially unmarked standard makes privileged ways of speaking appear mainstream, while marginalizing nonstandard dialects and speakers by placing them outside the mainstream. Specifically, NPR has utilized SAE, often characterized as "the broadcast standard" (Tamasi and Antieau 2015). But while SAE is characterized as having no regional accent (Lippi-Green 2011), the standard language uses the written language as its model but also draws from the spoken language of the upper-middle class (Milroy and Milroy 2012).

By presuming that network speakers have no distinct accent, NPR can more easily claim that it serves the American public, writ large. However, the very process of selecting a single way of speaking to serve as the national standard raises important political and moral considerations. Therefore, we must account for the language ideologies that are embedded within everyday broadcast practices. According to Irvine and Gal (2000), judgments about language are inextricably linked to judgments about speakers. Stigmatized vernaculars are used to signify the undesirable moral, intellectual, or social characteristics of marginalized speakers, while forms of speech possessed by those with social and political power will be seen as credible and will be upheld by social and cultural institutions. Irvine and Gal further argue that linguistic ideology must be seen as a totalizing vision in which elements that do not fit seamlessly into its interpretive structure are policed, co-opted, transformed, or suppressed in public space.

NPR's pursuit of linguistic homogeneity raises the issue of what Milroy and Milroy refer to as a "standard language ideology," which begins with the assumption that monolingualism, rather than linguistic diversity, is the natural state of being. Furthermore, standard language ideology is predicated on the belief that standard forms of speech are considered to be more elegant and logical than others. Embedded within the concept of standard language ideology is class and racial difference, in which the notion of a universal and socially unmarked standard makes privileged ways of speaking appear mainstream, while speakers of nonstandard dialects are positioned outside the mainstream. Because diversity is the norm, linguistic standardization requires a system of coercion, what Foucault (1984) has called "the disciplining of discourse." Milroy and Milroy (2012) further argue that these disciplining processes are enacted in everyday practices and policed by a number of professionals working in education, publishing, and broadcasting.

Maintaining the Broadcast Standard

When asked to describe the NPR voice, a station president with whom I spoke characterized it this way: "it's like a wise friend that has a lot of information." The idea that NPR speaks to its audience like a close friend was a recurring trope within NPR discourses. In the network's retrospective, *This is NPR: The First Forty Years* (2010), host Susan Stamberg recalls her early discussions with Bill Siemering regarding the voice that NPR wanted to cultivate. "We want NPR to sound more relaxed. Conversational," she recalls Siemering saying. "We're going to talk to our listeners just the same way we talk to our friends—simply, naturally." Jonathan Kern, executive producer for training for NPR news, echoes this sentiment in *Sound Reporting: NPR Guide to Audio Journalism*, stating "We should communicate to that archetypal listener, much in the way we actually talk to our friends or family" (Kern 2008, 27).

While the practice of treating the listener as a close friend may seem intuitive, the nature of speaking will, in reality, depend on who that friend is, the social environment in which that friend learned to speak, as well as the situational context of the speech exchange. As sociolinguists argue, spoken language varies not only regionally but also according to the social groups to which the speaker belongs (Duranti 1997).

Based on my interviews and a review of NPR's internal documents, I found that NPR has deliberately cultivated a speaking style that is meant to appear conversational but that is, in fact, highly scripted. Despite its pretense of inclusivity and informality, the NPR speaking style may be better described as an *idealized dialect* (Johnson 2000), meaning that it is a linguistic style that is not really spoken anywhere, but instead is acquired through professional training. In this way, NPR follows in the tradition of other broadcasting networks that promote uniform ways of speaking. Like other broadcast news networks, NPR utilizes a form of spoken English that converges toward a uniform pattern that is divested of regional identity through the use of "neutralization techniques."

While NPR has followed in the broadcast tradition of utilizing SAE, it has simultaneously developed an on-air linguistic style that differentiates the network from its competitors within the media marketplace. In a blog posting, NPR social media strategist Andy Carvin (2011) describes the NPR voice this way: "We've always prided ourselves on a delivery that's thoughtful, deliberate and clear, while not being rushed, or academic for that matter. You also won't hear us using a lot of fancy, multi-syllable words that are more common in newspapers and magazines. We really try to write our scripts so they come across as spoken English, not written English."

As Carvin makes clear, NPR's conversational style is not necessarily a reflection of spontaneous delivery, but rather a highly disciplined practice that is meant to mimic spontaneity. This way of speaking, referred to by some industry insiders as "midwestern nice" (Margolick 2012), is designed to position NPR as a sensible alternative to the authoritative, highly rehearsed speaking styles employed by many broadcast news organizations, what one participant referred to as "the voice of God."

At the same time, NPR's conversational approach distinguishes it from the more rapid speaking style of commercial radio, what was referred to by another informant as "puking." As a news organization that competes with other news organizations for audiences, the ability to develop a unique and recognizable voice can be a competitive advantage, endowing NPR correspondents with what Bourdieu terms *linguistic capital* (1991), or the capacity to speak appropriately for particular markets. The more linguistic capital speakers possess, the more they are able exploit the system of differences to their advantage and thereby secure a profit of distinction.

My interviews reveal that NPR employs both formal and informal mechanisms for maintaining the NPR standard. Explicitly, NPR practitioners are directed to an "NPR Pronunciation Guide," which is maintained by NPR's librarians and includes the latest NPR-approved pronunciations (Memmott 2014). Additionally, NPR hosts an online resource called "NPR Training: Storytelling tips and best practices," (National Public Radio n.d.), which is produced by the NPR training team and is intended to encourage practitioners to produce stories that are consistent with NPR's production values. Therefore, Jensen's claim that NPR's on-air pronunciation is at the discretion of the correspondent is not quite accurate. Practitioners are encouraged to avoid regional pronunciations in order to ensure clear communications. Consider the direction provided by NPR training regarding how to pronounce city names: "If a well-known city has an Anglicized name, use it if English is the reporter or host's native language. Don't call the capital of France 'pahr-EE' or say 'mosk-VAH' instead of Moscow. In general, it's better to go with English names for major cities such as Montreal, Lisbon, Cairo, Copenhagen, Mexico City, Geneva and Jerusalem, to name a few" (Socolovsky 2019).

Such broadcasting protocols are designed to create a clear, consistent sound. Informally, however, NPR also works with a number of contributors, reporters, and hosts who have been professionally groomed in programs where they have learned to speak in the neutral tone of public radio. This means that, by the time those individuals find themselves at NPR, they have learned to modify their way of speaking to align with the NPR norm. This can present a challenge in recruitment of Latinxs who have access to multiple linguistic codes. Consequently, NPR has had to had to turn to alternative pipelines of talent and then train them on how to speak in the NPR standard.

During my interview with A. Martínez, host of KPCC's *Take Two*, we discussed his challenge in making the professional transition into public radio. Martínez reported that, when KPCC was searching for a Latinx host, the station had originally sought to recruit someone with broadcast news experience. However, the station had difficulty finding a Latinx host with experience either in public radio or in English-language broadcast news. This prompted the station to expand its search. Martínez joked about the process, stating, "they [KPCC] reached down to the sports radio rock, turned it over . . . turned that slimy rock over, and found somebody that could actually do [this job]." According to Martínez, he originally struggled to find a balance between meeting the NPR standard and retaining his own voice. Here, Martínez describes the transition:

> [In commercial radio] it was a little more free-flowing. So, I could get a little more excited. Show a little more emotion. . . . I was wondering if I'd even be able to even do this, because it seemed a little too controlled and I had grown up on commercial radio. So, one of the things when I came here that I struggled with is, do I try to sound like public radio? What I think that public radio is, which is this controlled, snooty, elevated way of speaking?

Martínez is describing the linguistic hierarchy that exists within broadcasting, in which the speaking style of sports radio is seen as lowbrow relative to that of public radio. In an effort to adopt a new norm, Martínez told me that he worked closely with station producers to modify his speaking style in a way that would be consistent with other voices on the station, including his co-host, Madeline Brand.

Martínez' experience was not unique. NPR's on-air talent generally works closely with NPR editors, producers, and voice trainers. As Jensen (2019c) noted in her essay on NPR's linguistic practices, Mark Memmott serves as supervising senior editor, standards and practices, at the national level. In this role, Memmott is tasked with ensuring a consistent approach, including the use of language. These standards are communicated through formal training with new employees, and with a series of communications he calls "Memmos." However, similar forms of gatekeeping occur at the local level. My interviews with producers and station managers reveal that local correspondents are encouraged to sync up with the speaking styles of correspondents reporting from NPR headquarters in Washington, D.C. or Los Angeles. The ultimate goal is to create the perception of congruency between national and local productions. As one station manager whom I interviewed stated, "the listener should think that (both correspondents) are reporting from the same room."

NPR has also employed the use of voice coaches. Of these, NPR's David Candow had become the most infamous. Referred to within NPR as "the host whisperer" and the "Henry Higgins of public radio" (*Current* 2014), Candow is credited with helping to cultivate NPR's semi-informal speaking style. I spoke with Ray Suarez, former host of NPR's call-in show *Talk of the Nation*, who recalled working with Candow.

> In early 1998 or so, the then Vice President for News brought in a speech expert, David Candow. All of the show hosts had several workshops with David Candow. I was partnered with Robert Segal and we went to sessions with David Candow. Robert Segal has a gorgeous voice with beautiful diction. And I thought, he doesn't need training.

Candow's specific expertise was helping on-air talent to translate written language into something that sounded spontaneous and conversational, which can be a difficult task. As Candow reportedly said, "sounding natural and easy-going on the air can take years of practice" (Farhi 2008).

I had the opportunity to interview a voice coach who has worked extensively with a number of NPR's on-air correspondents, both at the national and local level. According to the participant, professionals working at local member stations need the most help, because they are less disciplined, stating, "at the local level, many reporters at our member stations are not quite up to [being on air]. They've learned bad habits." According to the participant, these "bad habits" include speaking too slowly or quickly, or not speaking clearly for broadcast purposes.

According to the participant, foreign-born journalists pose a particular challenge to broadcast producers. For example, the voice coach reported that she had worked with U.S. correspondents stationed in Iraq during the Gulf War, but had some difficulty working with Iraqi journalists, who did not speak in the NPR standard. While she described these journalists as an "invaluable resource" given their contributions to various stories, she also acknowledged that their way of speaking often prohibited them from delivering their material on-air. According to the participant: "This was one person, who was very smart and knowledgeable. But his accent was so thick that it was pretty darn impossible to understand him . . . there are some [journalists] where their accents are so thick that you just can't use them."

Here, the participant raises the important issue of accented English. When I asked her why there is a need for strict uniformity in speaking on NPR, she asserted that even subtle variations in speech have the potential to make a story harder to comprehend, stating, "if you pronounce a word that's different, people will be thrown off by that."

This belief that foreign pronunciation would be disruptive to the communications process was shared by several of the participants with whom I spoke. As Milroy and Milroy (2012) argue, such standardizing practices are based on the pretense of organizational efficiency, in which the primary goal is to develop a consistent form of communication that can travel over long distances and persist over long periods of time. However, this perspective reflects the normative belief that there is one and only one correct spoken way of speaking. Consequently, linguistic diversity is seen as the antithesis of clear communication. But there are clear racial implications. These practices ultimately serve the ideological purpose of establishing nonstandard forms of speech as deficient relative to the "correct" form.

NPR and the Sound of Whiteness

In 2015, NPR contributor Chenjerai Kumanyika wrote an article for *Transom* (2015), in which he described the implicit pressure for people of color to suppress their voices in order to find a home on public radio. In the essay, Kumanyika described a moment of realization, as he was preparing a story for air, when he became aware that he was altering his own speaking style to fit what he believed to be the NPR voice. According to Kumanyika:

> I realized that as I was speaking aloud I was also imagining someone else's voice saying my piece. The voice I was hearing and gradually beginning to imitate was something in-between the voice of Roman Mars and Sarah Koenig. Those two very different voices have many

complex and wonderful qualities. They also sound like white people. My natural voice—the voice that I use most often when I am most comfortable—doesn't sound like that. (Kumanyika 2015)

I had an opportunity to speak with Kumanyika after he wrote his essay. During our conversation, I asked him about the process by which he produces a piece that might air on public radio. Kumanyika described a two-step process, in which he begins by writing a script, for organization and clarity. The next step involves figuring out how to voice the script. This is the process by which written pieces are modified so that they appear conversational. According to Kumanyika:

You have to voice your piece. You have to read it . . . when you're reading, you kind of start orienting yourself to a thing that maybe you've heard. It's kind of like subconscious. I don't know if it's fully a conscious thing. But if I'm just reading it, I'm orienting it to the voices in my head. So, I pulled in the voices that I had heard, which are people that sound like NPR, you know, and basically that's not the way that I speak when I'm more relaxed.

I asked Kumanyika to elaborate on his argument that NPR's broadcast style is incongruent with ethnic speech. Kumanyika stated that, in his opinion, NPR's controlled way of speaking does not reflect some of the more emotive ways of speaking sometimes associated with ethnic speech, what he referred to as "bigness." Kumanyika told me that he loves hearing the range of voices on commercial radio, particularly those who work in hip hop and sports, stating, "you just hear folks be animated and you see all kinds of linguistic aesthetics come out."

Kumanyika also seemed to suggest that efforts to control speech were both implicit and explicit. Implicitly, he acknowledged his own inclination to model his scripts on NPR pieces that he has heard before. Because there are so few people of color on NPR, says Kumanyika, the exemplars are almost entirely voiced by white hosts and correspondents. But Kumanyika also described more active involvement by producers to control voice. According to Kumanyika:

If I'm telling the story of something that's shocking and crazy on NPR and I come into the studio like that, they're probably going to say to me ... particularly if you're working for like *All Things Considered*, they might just be like, "can you try that again?" Or "we want to go in this direction." There might be other language they might use, but clearly, you don't hear anybody [speak that way].

According to Kumanyika, it wasn't always that way on NPR. He believes that, when the network first started, the staff was much more willing to experiment with format and with voice. He believes that the network has become more conservative over time, stating: "When you look at the original NPR, like when you hear the first *All Things Considered*, it sounds like they was in there with drugs or something. I think the first voice was like a Black nurse from somewhere. It really was a lot different than what it ultimately became."

As Kumanyika suggested, I listened to the very first episode of *All Things Considered* (ATC), which debuted on May 3, 1971. Sure enough, the very first voice, after host Robert Connelly introduces the episode, is Janice, a Black mother and nurse, who is struggling with addiction. When I spoke with Bill Siemering, he acknowledged that this was intentional. "It was a statement I was making that the first voice you heard after the announcer was a Black nurse," Siemering told me. "Because I wanted to talk to primary sources. Not about a topic, but from the people who experienced it."

While the fact that the first Black voice heard on *ATC* is someone grappling with substance abuse is problematic in its own way, it is hard to deny that the larger episode reflects a broader range of voices. Much of that first *ATC* episode is dominated by coverage of the protests in Washington, D.C. against the Vietnam War. Rather than have correspondents mediate the experience, Connelly sets up the piece stating that NPR will take listeners directly to the event, where they can "get the feel, the texture, of the sort of day it's been, through a mix of sounds and events" (National Public Radio 2017).

During the segment, which runs over twenty minutes long, journalist Jeff Kamen is engaging a variety of people impacted by the protests. The result is a segment filled with local voices, which are clearly regionally and ethnically marked. In my own conversation with Kamen, who describes himself as a middle-class Jewish kid from New York, he admitted that, when he first started in broadcasting, he adopted the "midwestern flat" accent, believing that his own way of speaking may have been seen as "unseemly" because of anti-Semitism within the broadcasting industry. Like Kumanyika, Kamen also thinks that NPR was willing to take risks in its early days, but fears that it has become more mainstream over the years. "As the institution became a bigger, richer, more dominant voice in America," he told me, "there may have been a tamping down of that edge."

Listening to that first episode, two things become clear. First, the segments on NPR's magazine programs have become much shorter. Second, there is a heavier reliance on expert sources. Both practices have pushed out speech that is regionally or ethnically marked. Former NPR public editor Elizabeth Jensen told me that she believes that tighter show clocks have pushed out a greater range of voices. "You don't want to have a 20-minute piece today," Jensen told me. "So that is a problem. Because you don't have people speaking for themselves, you have actualities being cut out. You have people [whose] perspectives are being summarized by the correspondent."

Overall, a key finding of this research is that the on-air voice that NPR cultivates and maintains ultimately situates whiteness at its center. As with the concept of whiteness itself, standard English is defined in terms of what it is not. As Milroy (2001) writes, "the standard of popular perception is what is left behind when all the non-standard varieties spoken by disparaged persons such as Valley Girls, Hillbillies, Southerners, New Yorkers, African Americans, Asians, Mexican Americans, Cubans, and Puerto Ricans are set aside" (Milroy 2001, 59).

However, the coded nature of standard language ideology makes discussions of diversity on NPR difficult to ascertain. Unlike television news, which is visual, broadcast radio relies on a set of signifiers that are less visible. This practice allows NPR to lay claim to diversity, while not having to make any meaningful changes to its broadcast standard. To illustrate, several of the participants with whom I spoke defended NPR's practices by pointing to Lourdes Garcia-Navarro, host of NPR's *Weekend Edition*, and Audie Cornish, host of *All Things Considered*, as exemplars of NPR's commitment to diversity. However, both perform in SAE, which means that their way of speaking does not readily reveal their ethnic identities. As one participant told me, "many listeners are surprised to find out that Audie Cornish is Black."

This is not to say that Garcia-Navarro and Cornish are not speaking in ways that come naturally to them, or that there is no linguistic diversity within the Latinx and African American communities. This is to say, however, that both speak in a style that is highly congruent with the NPR standard. Overall, the ideal NPR Latinx is one who is, more or less, accent-less, and while NPR favors Latinx speakers who are proficient in SAE, there is a conspicuous absence of NPR correspondents who possess recognizable Latino accents. When they are represented, both Latino-accented English and African American Vernacular English (AAVE) appeared to be restricted to limited spaces within the programming schedule. For example, some linguistic variation is permitted in NPR's diversity-related programming, such as *Code Switch* and *Alt.Latino*, in which the hosts and actualities commonly engage in the practice of code-switching. However, I found that the hosts of NPR's flagship programs, *Morning Edition* and *All Things Considered*, maintain a strict commitment to the network standard.

That NPR would be protective of its flagship news properties is not necessarily surprising. Network newscasters and commentators commonly train themselves to speak a form of English that is devoid of any regionalisms. According to Bell (1983), such practices serve important symbolic needs. Broadcast speech must be seen as a prestige standard, due to the importance attributed to its subject matter. The news claims to represent authoritative information about national and international affairs and is regarded by audiences to be of social importance.

This means that NPR excludes forms of speech that may be perceived as undermining the prestige of its flagship news programs. These broadcast practices have particular implications for Latinx speech, which remains highly stigmatized within the United States, including the use of Spanish, code-switching, and accented English. According to Rosa (2016), the marginalization of Spanish in the United States is linked to the outsider status of many Latinxs, despite their longstanding presence in the United States. Urciuoli (1996) further argues that Latinx speech has become a signifier of the devalued position that Latinxs occupy within U.S. society, where "speaking with an accent" or speaking in "broken English" are positioned as inferior ways of speaking relative to "good English."

Here, Garcia-Navarro's presence on NPR requires further consideration. During my interviews, Garcia-Navarro was continuously brought up as an exemplar of NPR's tolerance of Latinx speech, but it's important to consider her case more fully. Garcia-Navarro was born in London and raised in Miami by parents that are Cuban and Panamanian. She is highly educated, having studied international relations at Georgetown University and having later received her MA in journalism from City University London.

Garcia-Navarro began her career working as a freelance journalist for the BBC World Service and Voice of America. As she mentioned on Twitter, Garcia-Navarro speaks Spanish and French, but she can capably perform in SAE, and she speaks almost exclusively in English in her role at *Weekend Edition*. Aside from pronouncing her name and a select set of cities authentically, there are very few instances in which she reports in Spanish. It is Garcia-Navarro's ability to master the broadcast standard, not her access to multiple linguistic codes, that has helped her succeed at NPR.

NPR's move toward linguistic standardization has been an ongoing process. When the network canceled *Enfoque Nacional* in 1988, NPR essentially cemented its position as an English monolingual space. This reality was formally acknowledged almost thirty years later, when NPR revisited its decision to cancel the show. In an essay titled "NPR in Spanish: Approaching Content for A Bilingual Audience" (Pretsky 2017), NPR's public editor's office acknowledged that, because over 40 million U.S. Latinxs speak Spanish at home, NPR had an obligation to create Spanish-language programming. However, there are limits to NPR's support of Spanish. In the essay, NPR senior vice-president of programming and audience development Anya Grundmann essentially argued that Spanish may be fine in podcast form but would not find its place on *Morning Edition* and *All Things Considered*. "We're basically an English-language American broadcaster trying to bring in all perspectives," Grundmann was quoted as saying. "For that reason, it's unlikely NPR listeners will again hear news content entirely in Spanish on the radio as in the *Enfoque Nacional* days."

Because of NPR's unrelenting pursuit of English monolingualism, the network has become more conservative than other public radio networks within the United States. For example, *Radio Bilingüe* offers programming in languages other than English and Spanish, and includes content in Mixteca and Triqui, indigenous languages spoken by some of its listeners (Green 2010). Similarly, PRI's *The World* is more likely to have on-air correspondents who speak with accents that reflect their countries of origin. The network is more likely to include longer segments of untranslated speech on-air, and, when translated, attempts are made to match the speaking style of the translator with their source material.

The disparity between NPR and other public radio networks prompted journalist and professor Gisele Regatao to write a critical essay in the *Columbia Journalism Review* (2018). In the piece, Regatao recounts her attempt to pitch a story to NPR, only to be told by an editor that her piece would not air out of concern regarding her accent. Regatao, who speaks in a slight Brazilian accent, has voiced stories for PRI and WNYC, and was surprised by NPR's categorical rejection at the national level. In her essay, Regatao wrote that NPR's rejection of her piece reflected the network's failure to accurately represent the country. "Ultimately, accents reflect who belongs and who doesn't," Regatao wrote in her essay, "and what the voice of our country sounds like" (Regatao 2018).

I asked Regatao why she felt compelled to write this essay. She told me that she was interested in the power that national news organizations have to exclude entire communities, stating, "it's not just about me and my voice. It's about a large percentage of people that are not being represented." During our conversation, she told me that she wanted to investigate the issue from an academic standpoint:

> I am a scholar of journalism, and the fact that an editor of a major, national public radio news organization was telling me, in an e-mail, that one of the reasons that my piece was not airing was because of my accent, I felt it was a subject that I needed to write about it. I really wanted to understand why people don't like accents. . . . I felt it was an issue that we needed to talk about.

I asked Regatao whether she thought these kinds of exclusionary practices were unique to NPR, or if she thought it was representative of public radio in general. During our conversation, she recounted a very different experience with PRI's *The World*, shortly after being rejected by NPR. She was approached by producers at PRI to develop a piece on blues artist Robert Johnson (2019) as part of their *American Icons* series. According to Regatao, she was self-conscious of her accent after her experience with NPR. Here, she recounts the experience:

> We had a meeting, and while in the meeting, this series is an audio documentary, and they're beautiful stories. But sometimes the journalist voices the story, and sometimes it's the host. So, in the meeting I was with the editor and the executive producer and I asked them, "do you want me to voice the story?" Because I know that sometimes the host voices the story. I was already anticipating that they were going to say the host. And they said, "no we want you to voice the story." And I said, "I want to make sure, because I have an accent." And they joked, "what accent?"

According to Regatao, these experiences reflect two fundamentally different perspectives on language. In her words, "just as one national show was commissioning me to do a long piece, another news organization was saying my accent was the reason my piece didn't air. It was quite interesting." This experience, of course, has left Regatao somewhat apprehensive about working with NPR in the future. As she told me, "I haven't pitched NPR since and I don't intend to. Nor do I need to."

Disciplining the Latinx Voice

In response to NPR's aggressive standardizing practices, Latinx speakers have learned to cultivate some ways of speaking while suppressing others. In his essay on NPR's approach to language, former NPR public editor Jeffrey Dvorkin (2005b) wrote that some NPR correspondents have been told to suppress their regional accents, citing Kentucky-born Noah Adams, a longtime NPR host and reporter, as an example. According to Dvorkin, Adams was told to "lose his accent" if he wanted to succeed in news radio. Dvorkin further stated that homogenizing practices are particularly acute with Black and Latino journalists, writing that: "For many journalists of color in public radio—African Americans and Hispanics—similar pressures exist, but with racial pressures. Some tell me that they are urged to sound less like themselves and more like the NPR 'standard,' whatever that is."

Several Latinx participants reported that, when they veer away from SAE, their speech can be subject to scrutiny, including the pronunciation of their own names, affirming Urcuioli's point that the outer sphere is a space of imposed order in which ethnic speech is tightly controlled. For example, I had a chance to speak with *Latino USA*'s María Hinojosa in her office in Harlem. Hinojosa has had a successful career in public radio, building an impressive personal brand, but also establishing Futuro Media, an independent, nonprofit newsroom. But Hinojosa's journey has not been easy. During our interview, Hinojosa described her first time reading a script that she had written and making the conscious decision to pronounce her own name authentically: "Being the first Latina correspondent at NPR, I was deciding who I was going to be. I can't say that I spent hours and hours and hours thinking about it. It was more like, oh my God, I have my first script. It read 'for NPR from Washington, D.C., I'm María Hinojosa [reading her name in an Anglicized voice].' Nah, 'NPR from Washington, D.C., I'm María Hinojosa [Spanish pronunciation].'"

According to Hinojosa, the decision to pronounce her own name authentically seemed natural, but she admits that it has come with critique, not only from journalists, but also from members of the listening public: "And that was seen as a hugely political issue. People loved it or they hated it. Letters would be written to NPR. Oh Sylvia Poggioli. Oh, that's beautiful. But letters would come to me, can you tell her to speak English and Americanize her name."

Hinojosa is referring to Sylvia Poggioli, NPR's senior European correspondent based out of Italy. Hinojosa's suggestion that listeners may respond well to an Italian accent, but that they are not as forgiving with Latinx accents, is also consistent with Jensen's observation that Latinx speakers receive a disproportionate amount of criticism from NPR listeners.

The Latinx journalists with whom I spoke generally had the ability to code-switch, meaning that they could report capably in SAE, but were proficient in Spanish. For example, I had a chance to interview

Adolfo Guzmán López, education correspondent at KPCC. Like Garcia-Navarro, Guzmán López reports in the NPR standard, but chooses to pronounce his name in Spanish. And like Garcia-Navarro, this practice has elicited responses from the listening public. Guzmán López's on-air pronunciation was even a point of humor in an episode of *The Simpsons*, in which a school administrator attempts to pronounce the names of a character, Isabel Adolfo-Guzmán López Gutiérrez, authentically (Guzmán López 2014).

While Guzmán López is bilingual, much of his work for NPR is in English. He attributed his proficiency in SAE to his early education experiences, stating "I went to predominantly white, middle class schools all through elementary, and junior high, and high school. Which is one of the reasons that I don't have an accent, I think, in English." A. Martínez similarly expressed a belief that his job at NPR was reliant on the fact that he did not speak with a recognizable accent, stating, "and the thing is, I don't even have an accent. I have a California accent, which has no accent. It sounds like being kind of neutral."

By describing their early childhood experiences, both Guzmán López and Martínez are expressing an implicit awareness of what Bourdieu calls *linguistic habitus* (1991), or dispositions about language acquired in the course of learning to speak. One's linguistic habitus is inscribed on the body and is manifested in the form of accents, intonations, and articulatory styles. But, as Bourdieu argues, success in any given field is dependent on the congruency between one's linguistic habitus and the field they occupy. Therefore, Latinx practitioners who have a command of SAE are structurally at an advantage relative to those whose way of speaking is seen as incongruent, or deviant.

However, I found that, within the context of NPR, the linguistic range of some Latinx speakers is devalued. The Latinx participants with whom I spoke had, to varying degrees, access to multiple linguistic codes, which included proficiency in SAE, Spanish, and code-switching. But this kind of linguistic range is devalued in the context of NPR, supporting Rosa's (2016) point that the linguistic dexterity of U.S. Latinxs is perpetually devalued in the context of U.S. monolingual society, where English is positioned as the language that best represents the nation. Rosa further argues that, in this context, Latinx speech is seen as a linguistic problem that needs fixing.

This puts NPR at odds with some commercial radio stations that benefit from linguistic diversity. Within the commercial media industry, for-profit stations are employing hosts who represent the audiences they wish to pursue. However, NPR's suppression of the range of Latinx speech inadvertently undermines its own strategic goal of reaching Latinx listeners. As its core audience grows older, NPR will need to pursue a younger, and therefore more ethnically diverse, listener, and language should play an essential role in helping them to enact this goal. As Del Valle argues, proficiency in Spanish can serve as an asset by enabling producers to reach new markets. Commercial stations have been at the forefront of this experimentation. In his own work on the language ideologies that are embedded within television industry practices, Chávez (2015) has found that media outlets have increasingly experimented with code-switching and accented English in an effort to capitalize on the projected growth of Latinxs in the United States. Rather than capitalizing on the linguistic range of Latinxs, NPR has constrained Latinx speech so that it conforms to the broadcast standard, thereby extinguishing any markers of ethnic identity, with the exception of a few sanctioned spaces. This practice suggests that NPR has little interest in adapting new linguistic strategies in order to draw in Latinx listeners. Instead, it reflects their longstanding practice of reaching acculturated, English-dominant Latinxs. Within this particular professional space, it is presumed that Latinx speakers will modify their own voices in an effort to be heard on NPR. The result is that the linguistic capital associated with having access to multiple forms of speech has relatively less currency in the context of NPR.

The Ghettoization of Latinx Speech

Latinx speakers who do not, or cannot, comply with the NPR standard find their voices peripheralized on NPR's broadcasts, either limited to the cultural-specific programming, relegated to the digital realm, or excluded altogether. Latinx practitioners are not unaware of NPR's racializing practices. For example, A. Martínez expressed frustration with his station's claiming to reach out to new audiences, yet adhering to a speaking style that favors the educated elite:

If there's a problem that public radio has had and claims to try to fix about itself is that it wants to be relatable to the everyday person. Well the everyday person isn't someone that has a master's degree. There are more people that have bachelor's degrees, or community college degrees, or no degrees than the other. So how do you attract them without alienating the core, which is highly educated, white, 40–60 men? So how do you do that? That's one of the things when I came here that I struggled with.

Here, Martínez is expressing the dilemma faced by several participants I interviewed: How do you reach out to Latinxs without isolating NPR's core audience? The station managers and news directors with whom I spoke were candid in their assessment that it is difficult to include nonstandard forms of speech without isolating their current audience. These exclusionary practices were particularly evident at stations located in cities in which Latinxs represented a significant part of the population. When I asked Guzmán López about how his station captures the stories of the many Latinx residents who are not English proficient, he described the practice of heavily editing the audio testimonies of Spanish speakers: "In [this city], I've come across Spanish-speaking people with great stories. Really great stories. But here, I can only use a [sound] bite here or a [sound] bite there. If they are only Spanish-speakers, I'll have to paraphrase, or I used to have someone else voice the translation."

When I asked Guzmán López to discuss why his station felt the need to limit Spanish in this way, he reported that there is an implicit understanding that one should not isolate the station's core listeners. Guzmán López continued: "We're not going to let them [Spanish-speakers] speak for fifteen seconds. Because the general listener just doesn't understand, and they'll tune out. But let them speak in their language for five seconds or so, and then as the reporter I'll come in and paraphrase."

Here, the "general listener" is understood to be English monolingual, which predetermines which kinds of content will be seen as acceptable and which will be seen as problematic. Because stations are afraid that Spanish-language testimony has the potential to exclude or divide listeners, news directors and station managers are conservative in their approach.

A convenient solution to this dilemma has been to relegate some forms of speech to the digital realm, where it is less likely to interfere with the on-air broadcast experience. For instance, in 2016, NPR began to distribute and promote *Radio Ambulante*, a Spanish-language narrative podcast that, according to NPR, is intended to tell the stories of Spanish speakers across the Americas. The podcast is marketed through NPR's primary website, as well as the NPR One app and iTunes. The solution of providing unique spaces for Latinx speech may enable NPR to claim that it offers diversity programming, but there is a ghettoizing effect. By confining Latinx speech to limited spaces within NPR, the organization is engaging in a form of language-based subordination in which standard English is deemed the authoritative language, appropriate for serious news, while trivializing Latinx speech, considered appropriate only for niche audiences.

I also found that Spanish-language content is often redirected to alternative outlets, both public and commercial, that are more open to broadcasting Spanish-language content. This was a practice that NPR first publicly signaled when NPR executives argued that *Enfoque Nacional* was more appropriate for commercial Spanish-language stations. During my conversation with Guzmán López, he discussed "having an itch" to produce Spanish-language stories that are focused on native Spanish speakers. Despite working in a city that has a high presence of Latinxs, Guzmán López reported that KPCC has never aired his Spanish-language work. In response, he has built relationships with Spanish-language stations located along the U.S.-Mexico border. Similarly, Radio Bilingüe, a community-based radio network for the nation's Spanish-speaking community, has become another destination for stories that are produced in Spanish.

I had a similar discussion with a former producer at *StoryCorps*, a weekly series featured on NPR, which focuses on the lived experiences of everyday Americans. According to the *StoryCorps* website, the goal of the organization is "to preserve and share humanity's stories in order to build connections between people and create a more just and compassionate world."

In an effort to capture a wide variety of stories, the producer at *StoryCorps* stated that the production team begins by operating in cities that are demographically diverse, trying to understand the demographics within that city, and then, to the degree possible, replicating them—what the producer called "taking a snapshot" of the city. Despite its promise that "understanding that everyone's story matters," however, I found there to be a conspicuous absence of stories in Spanish on NPR, which appears to be a striking omission, given that roughly a third of Latinxs are Spanish-dominant.

When I asked the producer how *StoryCorps* dealt with the testimonies of Spanish-dominant Latinxs, she reported that they hire individuals who can communicate in Spanish and can help record their stories, what she called "facilitators." These stories are made available online through a resource called *StoryCorps Historias*. While they had resolved this on the production side, the participant admitted that they faced challenges on the distribution side. After all, NPR is broadcast in English and will not air Spanish-language content, thereby excluding the stories of these individuals. With NPR as a non-option, the producers looked to Radio Bilingüe as an alternative outlet.

The ways in which Latinx speech is suppressed on-air was evident during my interviews at a station in a large, metropolitan market with a small but growing Latinx population. This was one of the rare instances in which I encountered an NPR member station that had hired a Latinx correspondent who was Spanish-dominant and, therefore, did not report in SAE. The station's news director was optimistic that the reporter could begin to build bridges in the area's growing Latinx community by focusing on social issues, including education. According to the news director: "We've hired someone from [Spanish-language media] and her Spanish work ends up in social media. We have a web page in Spanish. In the future, could we become a partner to Spanish language radio stations and give them content?"

The station manager's testimony reflects the tension between reaching out to Latinx listeners while ensuring a consistent experience with NPR's core audience. Whether driven by a genuine interest in serving the Latinx community or a practical desire to expand its reach, the station has invested resources in a Spanish-proficient journalist. At the same time, the station's managers ensure that the work of this journalist will not be heard on a wide scale. As the participant states, the Latinx reporter's stories are promoted through social media and can also be found on the station's website under a separate Spanish tab.

While the news director feels that her work might be appropriate for Spanish-language radio stations, he makes clear that it is inappropriate for the station's on-air broadcast. For example, the reporter has worked on a series focused on education, which features the stories of Latinx children. These stories provide insight into the ways in which Latinx children negotiate a variety of personal, educational, and political pressures. One of the stories provides insight into how Latinx children are affected by federal immigration policy. Another story addresses how Latinx children serve as linguistic intermediaries between their parents and social institutions within the United States, including the education system. Because these stories are produced in Spanish, their distribution is restricted from the station's on-air broadcasts.

Because the journalist speaks in accented English, she also receives few opportunities to report in English. When I asked why this reporter does not have more of an on-air presence, the news director reported that he has received pushback from the audience: "Sometimes our listeners have a hard time understanding her. And we'll get e-mails or people calling in. 'Why are you putting her on the air, we can't understand her?' But it's important, we have Spanish speaking listeners, and her English will get better."

In a follow-up interview with this journalist, I inquired about the process by which her work is developed and produced. For example, who assumes responsibility for editing her work, since she is the only employee at her station who is a native Spanish speaker? The informant acknowledged that there is currently nobody at the station who is able to edit her Spanish-language scripts. However, she told me that, for larger projects, the station has hired the services of a Spanish-language instructor at a local university—someone who is not Latinx.

While this practice is meant to ensure that the reporter's scripts adhere to proper syntax and grammar, it does not allow for editorial feedback that is standard practice for refining audio storytelling. Consequently, the reporter lacks the necessary feedback and collaboration with her colleagues that will help her move forward professionally. Ultimately, the journalist's case illustrates a number of important insights. First, the audience's reaction to Spanish, as well as the station's response, demonstrate the various ways in which forms of speech that are considered deviant are policed. Second, the lack of investment in organizational resources will impede the journalist's ability meaningfully to develop Spanish-language content, thereby inhibiting her professional development. Finally, the journalist's relegation to the digital realm illustrates how alternative forms of Latinx speech are restricted within the programming schedule. These issues were not lost on the journalist. During our conversation, she expressed a sober awareness that her accented English was a professional liability at NPR, stating "I wish I had taken an accent reduction class. Sometimes I'll visit with students who want to work in radio. And I tell them, take a class."

Linguistic Standardization and Its Constraints on Civic Discourse

Calls for developing uniform ways of speaking are not new, nor are they limited to the broadcasting industry. Discourses regarding civic engagement within the United States have long reflected concerns about linguistic competency. As a "nation of immigrants," the country has been characterized by linguistic plurality, yet there have been constant calls to ground civic discourses in a single, shared language. These discussions on which kind of speech is best suited for political participation date back to the early days of the republic. For example, John Adams's vision for an American Academy was based on the belief that governance required a common, but uniquely American, way of speaking (Tamasi and Antieau, 2015).

The capacity for individuals to engage in civic discourses is tied to what Dell Hymes (1972) calls *communicative competence,* or the capacity of persons to recognize and select the language variety appropriate to the occasion. According to Dell Hymes, different speakers possess different *linguistic repertoires* and learn to select from this repertoire in order to meet a number of communicative needs.

In his own work in civic discourse, Habermas (1970) calls for specific linguistic conditions that might ensure that all citizens would have the opportunity to be heard on equal terms, what he calls an "ideal speech situation." Under such conditions, participants would be able to evaluate each other's assertions solely on the basis of reason and evidence. In such a speech situation, participants would be completely free of any physical and psychological coercion. However, Habermas's notion of the public sphere does not account for the ways in which power is enacted in everyday speech. After all, not all speakers can engage in civic discourse on equal footing. In his work on language and power, Bourdieu (1991) argues that every speech encounter, both formal and informal, reflects hierarchies at play in the larger social space. In any society, specific social-historical conditions endow the dominant language with the status of the sole legitimate or official language.

Originally, NPR was meant to address these concerns by engaging a greater variety of speakers, but, over time, the network has gravitated toward linguistic standardization. Because NPR begins with monolingualism as its starting point, the linguistic dexterity of Latinx speakers is seen as a deficiency rather than a strength. A key finding in this research is that NPR has cultivated a unique voice that is meant to distinguish the network in a competitive media marketplace. But I found that the construction of this voice is not arbitrary. Instead, these practices reflect racial ideologies that are meant to suppress Latinxs in white public space. In Garbes's (2017) words, it is a norming process that "logically places all traditionally nonwhite voices into an alternative space of the organization rather than at the base, despite its mission to serve all American publics."

These findings extend Loviglio's (2008) argument that NPR is a cultural institution that is constituted largely by sound, and that by attending to vocal elements such as pitch, intonation, loudness, and accent we may better understand the ideological meanings that are communicated through the voices associated with the network. Because NPR's ideal listener is one who is English monolingual, Latinx speech must be made digestible and comprehensible, a process that necessarily involves excluding those who speak in Spanish or with accented English. This kind of "cultural work" (Loviglio 2008), performed by NPR, ensures that speakers of SAE (hosts, actualities, expert sources, and so forth) are privileged, while at the same time removing the markers of ethnic identity that might result in a sense of belonging for Latinx listeners.

LATINO USA

This was the reason I left National Public Radio to start the program *Latino USA* in the early 1990s. As an editor on the network's National Desk, it was frustratingly apparent that even some of my very talented and educated colleagues lacked an understanding of the complexities of the Latino experience. Often, when Latino issues were even considered, the focus was on viewing Hispanics as "problem people" . . . I wanted to create a radio vehicle that would portray the whole of the Latino experience, its complexity and diversity, its beauty and its pain.

—MARÍA MARTIN, *CROSSING BORDERS* 2017

On May 18, 2015, José de Jesús Deniz-Sahagun, a thirty-one-year-old Mexican national, was admitted into an ICE detention center in Eloy, Arizona. Two days later, he was pronounced dead by asphyxiation. The death was ruled a suicide, but Deniz-Sahagun's treatment by immigration officials prompted a two-hundred-person hunger strike (Ingram 2015), public protests outside the detention center, and a Congressional call for investigation into the center's treatment of immigrants (Julia 2015).

For over a year, Deniz-Sahagun's death, and the protests that surrounded it, remained primarily a local story, but, in 2016, *Latino USA* aired a two-part docuseries titled *The Strange Death of José de Jesús*, produced by Fernanda Echávarri, Marlon Bishop, and María Hinojosa. At over ninety-five minutes long, the series paints a complicated portrait of one man's life. The *Strange Death of José de Jesús* can be devastating to listen to. Through interviews conducted with family members, recorded in Spanish and English, we learn of Deniz-Sahagun's life as a father, a son, a brother, and someone with struggles.

While the focus is on a single case, the series also speaks to the larger issues of U.S. immigration policy, mental health resources, and the role of private companies that run immigration detention centers. As part of their investigation, the *Latino USA* team took trips to Arizona, Nevada, and Mexico, where they conducted interviews with medical staff and social advocates, as well as detainees.

The story elevated the profile of *Latino USA*. In 2017, the show won the Robert F. Kennedy Human Rights Award, and, by 2020, the series was downloaded over 58,000 times. A special article and webpage for the series has received over 10,000 views (Futuro Media Group 2020). The episode was also featured on *Think*, a national call-in program produced by KERA out of north Texas. Telemundo Arizona integrated *Latino USA*'s reporting to produce a four-part video series titled *Celda 603: La Agonía de un Inmigrante* (Félix 2017).

NPR, which had not covered the original case in 2015, carried the story on its website, while a seven-minute version of the story was aired on *All Things Considered* (Hinojosa 2017). In July 2016, *Latino USA* partnered with *Radio Ambulante*, a Spanish-language podcast that is also distributed by NPR, to produce a thirty-two-minute, Spanish-language version of the story, which included an interview between Echávarri and a host, *Radio Ambulante*'s Daniel Alarcón (Alarcón 2016).

By that time, *Radio Ambulante* had itself produced a number of stories on immigration. That year, NPR's music-oriented podcast, *Alt. Latino*, also produced a show on its own podcast, critiquing Trump's family separation policy by showcasing the efforts of musicians across the globe to highlight the devasting human impact on Latinx families (Contreras 2017). Collectively, these three programs were engaging in a critique of U.S. immigration policy, which was profoundly impacting Latinx communities.

Latino USA, *Radio Ambulante*, and *Alt.Latino* are three Latinx-produced, national radio programs that have been supported by NPR. At their best, these programs can be important, profound, and sometimes oppositional. By telling the stories of Latinxs in their own words, they can serve as an important form of representation in a media landscape that can often ignore them. Furthermore, they have the potential to address important civic issues by drawing attention to social issues of importance to the Latinx community.

Over the next three chapters, I discuss these three programs, each a case study in navigating the public radio system. In doing so, I draw upon critical media industry studies (Havens, Lotz, and Tinic 2009), which account for the ways in which individual media practitioners navigate the hegemonic practices of dominant media institutions. As with critical political economy, research in this area is concerned with the significant role that media institutions play in creating cultural meaning (Hardy 2014). However, Havens, Lotz, and Tinic argue that there is a need also to account for the complexities and contradictions that exist within media organizations. Therefore, Latinx media producers must not be seen simply as passive agents in the larger process of cultural production. Instead, we must account for the many creative ways in which Latinx practitioners, working with and within NPR, challenge representational practices.

From this perspective, radio programs like *Latino USA*, *Radio Ambulante*, and *Alt.Latino* exemplify the potential for Latinx practitioners within public radio to negotiate, and at times even subvert, the constraints imposed by NPR for their own purposes. However, a review of NPR's programming over the past fifty years suggests that these programs are highly exceptional cases. Therefore, it becomes clear that not just anybody can start a Latinx-oriented radio program on NPR. Rather, it takes a specific kind of Latinx practitioner, under a specific set of conditions, to establish a viable program on public radio. To understand these conditions, I also look at Bourdieu's general theory of practice (1977), which means looking at journalism as a social field that operates according to its own rules and logics of practice.

Within each field, agents have, to varying degrees, an embodied, natural feel for the game, what Bourdieu refers to as *habitus*. Furthermore, the practices particular to a given field favor some agents over others, and it is the congruency between their habitus and the field they occupy that determines the ability for these agents to succeed. But Schultz (2007) also describes a *professional habitus*, which is the mastering of a specific, professional game in a specific, professional field. For example, possessing journalistic habitus implies understanding the journalistic game and being able to master the rules of the same game.

According to Bourdieu, fields can be both conservative and dynamic. Fields are conservative in that those who hold dominant positions within the field will seek to maintain the status quo, thereby contributing to simple reproduction of the field. Journalists have been socialized to accept the rules, routines, and roles of journalism as matters of common sense. In other words, members of the journalistic field generally accept the assumptions of the field, so disruption from within is limited.

However, fields are marked by change, and newcomers are continuously entering the field, seeking to add some small difference. As these newcomers enter the field, they often bring with them their own dispositions and practices, which clash with the prevailing norms of production. Because those that hold dominant positions within the field are interested in maintaining the status quo, Bourdieu argues that these newcomers cannot succeed without the help of external disruptions, which include technological developments that change the nature of production, demographic shifts that create new affinities with these new producers, and political breaks.

Established at different points in time, *Latino USA*, *Radio Ambulante*, and *Alt.Latino* were made possible by a number of economic, demographic, and technological disruptions that have impacted public radio. To varying degrees, each of these programs demonstrates the ways in which Latinx practitioners have been able to exploit these disruptions in the public radio system to their advantage. Yet each show offers unique insights. They differ in how they imagine their ideal Latinx listeners and in the ways they use language to reach their intended audience. They also differ in the ways in which they utilize communications technologies to produce and distribute their content, as well as to seek collaborations and partnerships beyond the United States. Finally, the Latinx practitioners with whom I spoke have access to various forms of capital, which have been used to negotiate the power-laden relationship with NPR.

Latino USA and the Embodiment of Capital

I had the opportunity to interview the *Latino USA* team in Harlem, a part of New York that is home to a thriving Latinx community that includes a mix of residents of Puerto Rican, Dominican, Salvadoran, and Mexican heritage. Like Harlem itself, the journalists and producers that I encountered at *Latino USA* reflect the diversity within the Latinx community, which is rare within the news industry. According to the Pew Research Center, only 23 percent of newsroom staff are non-white (Grieco 2018). By contrast, *Latino USA's* media staff is 78 percent non-white (Futuro Media Group 2017).

Latino USA

According to Nadia Reiman, who was a senior editor at *Latino USA* at the time, this kind of diversity helps the team to accomplish their goal of representing the range of the Latinx experience. "We are a cornucopia of experiences," she told me. "We are many." Reiman prefers to produce, in her words, "stories from Latinos rather than stories about Latinos," meaning that, rather than having Latinxs be passive objects of study, she would like to see Latinxs as active agents in describing their own experiences. Furthermore, she believes that, having a primarily Latinx newsroom, they are better equipped to tell a wider range of stories about the Latinx community.

Leading the team is María Hinojosa, who has, over time, become inextricably linked to the *Latino USA* brand. Hinojosa has been the host of the show since it first launched in 1993 and has navigated its course over the past several decades. A veteran journalist, Hinojosa embodies the kind of capital necessary to lead a team of journalists, negotiate even-handedly with NPR, work with foundations to secure funding, and build community with listeners. As one staffer told me, Hinojosa is central to the ethos of the organization, stating "María is María and her spirit kind of infiltrates everything that we do."

I had a chance to witness Hinojosa's deftness at navigating a variety of publics firsthand, when she gave a public talk at the University of Oregon, in front of a crowd of several hundred. For an hour and a half, Hinojosa engaged listeners with stories about her childhood and experiences from the field, and offered a pointed critique of journalism and its relationship to the Latinx community (Oregon Humanities Center 2017). She is a charismatic speaker and easily connects with her listeners. Hinojosa's ability to navigate these various social situations is likely a product of her personal and professional habitus; dispositions are acquired through a gradual process of inculcation in which early childhood experiences are important (Bourdieu 1977).

This is also true of her linguistic habitus. Hinojosa is proficient in the broadcast standard favored by NPR, but she can also code-switch strategically. While her decision to pronounce certain words in Spanish has drawn negative attention from some listeners, her way of speaking has also served as a form of linguistic capital. Her access to both Spanish and English allows her greater range than her English-monolingual counterparts, which has been professionally advantageous throughout her career.

Born in Mexico City, Hinojosa immigrated to the United States when she was eighteen months old (Martínez Wood 2007) and grew up in a solidly middle-class family. She later attended Barnard College, a private liberal arts college in Manhattan, which is where she had her first foray into radio, taking on the Wednesday-night slot at WKCR, Columbia University's radio station (Gildea and Zuckerman 2020), while still a student at Barnard.

With the exception of a few roles in commercial media, Hinojosa is, for the most part, a product of the public radio system, which means that she has also acquired the requisite professional habitus, or, rather, an understanding of the journalistic game, and is able to master the rules of public radio. She began working as an intern on NPR's *All Things Considered* in Washington, D.C., where she was selected by Susan Stamberg (Gildea and Zuckerman 2020), one of NPR's "founding mothers" (Hajek 2015) and another graduate of Barnard College. That experience turned into a full-time position as production assistant. Later, Hinojosa moved to the West Coast, where she worked as an associate producer for NPR's ill-fated *Enfoque Nacional*, but moved back to NPR headquarters, becoming NPR's first Latina correspondent. In addition to her work on public radio, Hinojosa has also worked on the PBS television newsmagazine *NOW*, and has had opportunities working in commercial media, including CNN, *Time* magazine, and CBS Radio.

Over the course of her career, Hinojosa has been recognized with a number of the industry's most prominent awards, including the Robert F. Kennedy Award, the Peabody, the Edward R. Murrow Award, and the John Chancellor Award for Excellence in Journalism. She has also won several Emmys for her work. In 2019, Hinojosa accepted a three-year appointment as journalist in residence at her alma mater, Barnard College.

Hinojosa has, in short, what Schultz (2007) would describe as journalistic capital, a form of symbolic capital within the journalistic field that is closely connected to the concept of peer recognition. As Schultz argues, having a lot of journalistic capital means having a lot of respect from journalistic colleagues and having a good position internally in the journalistic hierarchy. According to Schultz, journalistic capital can be changed into other forms of capital and can be found in the small details of everyday newsroom practice.

In Hinojosa's case, she has been able to parlay this kind of capital into assets that benefit the show. For instance, she has been instrumental in gaining high-profile interviews. This was exemplified during the 2020 election season, when *Latino USA* was able to secure interviews with Democratic presidential front-runners, including Elizabeth Warren (Hinojosa 2020a), Bernie Sanders (Hinojosa 2019a), Pete Buttigieg (Hinojosa 2020b), Julián Castro (Hinojosa 2019b), and Corey Booker (Hinojosa 2019c). At the same time, Hinojosa also possesses what Thornton (1996) terms "sub-cultural capital," or knowledge and commodities acquired by members of a subculture that raise their status and differentiate them from members of other groups.

In Hinojosa's case, she has been able to distinguish herself and her program within a highly competitive journalistic marketplace by focusing on Latinx-oriented issues. She has been named one of the one hundred most influential Latinos in the United States by *Hispanic Business* magazine, and she has been celebrated with Latino-specific awards, including the National Council of la Raza's Ruben Salazar Award and the National Association of Hispanic Journalists Radio Award. In the process, she has gained access to prominent figures within the Latinx community, including Supreme Court Justice Sonia Sotomayor (2018), author Isabel Allende (2015), and activist Dolores Huerta.

Hinojosa's credibility within the Latinx community was particularly evident in an episode titled "Digging into 'American Dirt'" (Hinojosa 2020c), which focused on the controversy surrounding *American Dirt*, a novel whose publication drew national attention to the racializing practices within the publishing industry. Essentially a narco-adventure, the book had been promoted by the publisher as important literary fare and was endorsed by notable authors and celebrities, including publishing darling Don Winslow, who went so far as to tout the book as "the Grapes of Wrath of Our Times."

The novel, however, was heavily criticized for its stereotypical treatment of Latinxs by its white author, Jeanine Cummins, who received a substantial advance for the book (Olivas 2020). Latinxs took to social media to decry the book as well as a publishing industry that shuts out Latinx authors while promoting white authors to tell their stories. In its report on the controversy, Hinojosa and her team interviewed several key players involved in the controversy, including Cummins, but also Myriam Gurba (2019), who wrote a defiant critique of the novel that circulated widely on social media, and author Luis Alberto Urrea, whose work Cummins had been accused of directly appropriating. The centerpiece of the podcast, however, was an exclusive interview with esteemed author Sandra Cisneros, who had endorsed the book and has continued to support it publicly.

The product of this reporting is a story that sheds light on the publishing industry and its practices, which inhibit its ability to include the voices of more authors. The podcast was cited in a number of national publications, including the *Los Angeles Times* (Pineda 2020). Access to this particular set of actors would not have been possible without Hinojosa's credibility within the field. This was particularly true of Cisneros, with whom Hinojosa has had a longstanding relationship. Getting to this point, however, has been a long time in the making. *Latino USA* may have started off as a small, upstart radio program, but it has, over time, become a highly entrepreneurial organization with significant political influence.

A Brief History of *Latino USA*

While Hinojosa's name has become synonymous with *Latino USA*, the program began as a collaboration among a number of professionals working in education, college radio, and private foundations. Three years before *Latino USA* launched, Yolanda Felton, a development specialist for University of Texas radio station KUT, approached John Hanson, executive producer for the Longhorn Radio Network (LRN), about an idea to produce a radio program that would serve the Mexican American audience. The group then approached Gilberto Cardenas, who at the time was director of the university's Center for Mexican American Studies (CMAS), to produce the program (Tovares 2000).

Given the working title of *A Tiempo*, a proposal for funding for the program was submitted to the Ford Foundation. In an effort to assess the program's feasibility, Christine Cuevas, a program officer at the Ford Foundation, requested more details about the target audience, the proposed format, production costs, and the strategy for executing these plans. According to Tovares (2000), Cuevas saw the benefits of working with CMAS as a producer of the program. The center had support from the university and access to Latinx scholars nationwide. Furthermore, because of its relationship with the University of Texas, *Latino USA* could benefit from a distribution platform that was already in place. The LRN included a number of affiliate stations as well as access to satellite communications, which facilitated the distribution of the program to stations in larger markets.

Cuevas also reached out María Martin, who, at the time, was a twenty-year veteran of public radio programming who specialized in Latinx issues. Martin had worked on NPR's short-lived *Latin File*, a daily fifteen-minute English language show, and was serving as NPR's Latino affairs editor. I had a chance to speak Martin, who recalled how she first became involved with the project:

> Basically, I got a package of stuff with a proposal that a friend at the Ford Foundation had received this proposal for a Latino program. And she [Cuevas] said, what do you think? . . . The programming that was coming out of the University of Texas was some ways ahead of its time. The *Mexican American Experience* and *In Black America*. But they were basically interview programs. And I know that Latinos at this point in public radio had matured to such an extent, that we could do much better. We could do what NPR is doing, but we could do it with authenticity and context.

Martin had hoped to use her experience at NPR to develop quality storytelling, but she wanted to avoid the mistakes she encountered with NPR's *Latin File*, which she has described "a tokenistic approach to programming that would eventually fail" (Martin 2020, 44). *Latino USA* could be different, since Latinxs would have editorial control. But first the team needed to overcome the hurdle of gaining support from both the CPB and NPR.

According to Tovares (2000), *Latino USA* benefited from the economic disruptions that were occurring in public radio. By the mid-eighties, NPR had pursued a shift in strategy due to cuts in federal funding for public broadcasting. Rather than serving primarily as a producer of content, NPR would act as a distributor of programs that were acquired from independent producers. These producers would incur the production costs, which would, in turn, give NPR the financial flexibility to keep producing its flagship programs, *All Things Con-*

sidered and *Morning Edition.* This opened up the door for independent producers to develop programs that could be distributed by NPR.

However, convincing NPR to support a Latinx-oriented radio program would be a challenge. The development of *Latino USA* was occurring during the time that NPR researcher David Giovannoni was advocating that NPR member stations avoid diversity programming. Between the AUDIENCE 88 and AUDIENCE 98 reports, station managers were being encouraged to adopt NPR's "strategy of transcendence," which would avoid appealing to specific ethnic groups.

"This was when the research in public radio was telling the system that nobody was listening to discrete programs," Martin told me. "And so all of those little openings that had happened in the 1970s, and early efforts to have diverse voices on were being closed, because they said that people listened to format streams, not to discrete programs."* According to Martin, Giovannoni seemed indifferent to how his strategy excluded Latinx listeners. "I spoke with him, and he said basically, 'when Latinos get their master's degrees, then they'll listen to public radio,'" Martin said. "I used to have big debates with him, but to no avail."

To demonstrate its appeal to a broad audience, the premise of the show was reworked. Rather than focusing specifically on Mexican Americans, as was originally envisioned, the team now pushed for the more broadly defined "Latino," which was being promoted by the Spanish-language media. The team also needed to clearly communicate that show would be produced in English. The cancelation of *Enfoque Nacional*, just five years before, signaled that NPR would not be friendly to a Spanish-language program, but it also meant that more journalists and contributors could be invited to participate in the show.

The team decided to abandon the show's original title, *A Tiempo*, since it did not clearly communicate to listeners and station managers that the program was in English (Tovares 2000). Martin told me that they also considered the name *Mosaic*, since the goal was to reflect the diversity of the Latinx community. However, the team felt that

---

*By 1974, NPR had created the Department of Specialized Audience Programs, which developed a number of minority-oriented programs. With NPR's strategy of transcendence, specialized programs were later discouraged.

Mosaic did not represent the program's nationwide reach. In the end, they settled on the name *Latino USA*. According to Martin, "*Latino USA* said what it was. It was like, we're here, and we're covering the whole country."

As the show moved closer to production, Martin stepped in as the show's executive producer and, in turn, hired María Hinojosa to serve as the program's host. Hinojosa was herself a veteran journalist with significant experience in public radio. Additionally, Martin reached out to Latinx journalists and contributors who had experience in public radio, including Mandalit del Barco, who would go on to have a longstanding career at NPR, and Felix Contreras, who would later go on to produce and host *Alt.Latino*. When discussing her vision for the show, Martin described the balancing act between creating a show that would reflect distinctly Latinx sensibilities, yet sound like an NPR product: "At first it was to be authentic, but to sound like NPR so we could get on. So my first step was like sound a little different, but mostly like NPR so that you can get on the station, and then establish yourself. And then begin to play with experimenting both in language and format and whatever."

When it first aired on May 5, 1993, *Latino USA* was formatted as a half-hour radio program that typically opened with a news segment of about five minutes in length, followed by one or two segments that may examine an issue relevant to the Latino community or profile someone in the Latino community (Tatum 2013). A log of the stories from *Latino USA* acts as a sort of a historical record of the U.S. Latinx experience over the past several decades. During that time, *Latino USA* has covered issues such as the death of civil rights activist Cesar Chávez, the impact of the North American Free Trade Agreement (NAFTA), and California's Proposition 187, which sought to deny undocumented immigrants access to health and education services.

Martin left the program in 2002, and Hinojosa was elevated to the position of managing editor while maintaining her duties as host (Tatum, 2013). In 2010, Hinojosa took over production of *Latino USA* from CMAS and founded Futuro Media Group, an independent, nonprofit media organization. This move involved expanding *Latino USA* to an hour-long format (Jensen 2013), which required additional investment in developing more material that would reflect the experience of Latinxs in the United States.

Latino USA has grown significantly over its history. By 2020, the program was being carried on 222 broadcast stations nationwide (Futuro Media Group 2020), over twice as many as when it first aired. Furthermore, the program has expanded from a purely terrestrial radio program to a multimedia experience, which includes a strong online component. The *Latino USA* experience now includes a website that hosts the *Latino USA* stories, but also relevant stories produced by the Associated Press and Latin American News Dispatch, a news organization based out of New York University. The show is currently available as a podcast, which has increased its overall distribution, averaging 90,000 downloads per week (Futuro Media Group 2020).

Despite this growth, the editorial focus has remained the same. *Latino USA* continues to focus on policy issues that affect Latinx communities, such as national elections, the U.S. census, and legislative processes. *Latino USA* also continues to produce programs that focus on the human impact of immigration policy. For example, *Latino USA* included a segment titled "Seeking Asylum, Seeking to Stay Together" (Hinojosa 2019d), which focused on the plight of LGBTQ migrants. In 2018, the program ran a segment called "A Child Lost in Translation," which focused on the plight of migrants who speak Q'anjob'al (Hinojosa 2019e).

In addition, the program continues to feature Latinx arts and culture, showcasing Latinx musicians, artists, actors, and writers. This includes a mix of mainstream artists, such as actor Danny Trejo, musician José Feliciano, and rapper Pitbull, along with artists who have less mainstream appeal, such as Ranchera singer Ángela Aguilar (2019) and Rosalía Vila Tobella, a Spanish flamenco singer and songwriter.

I asked the *Latino USA* team about how they saw their mission relative to other news organizations within the larger media landscape. According to Marlon Bishop, who was a senior editor on the show, they cannot duplicate the work of better-funded organizations, like the *New York Times*, that have the resources to deliver breaking news and engage in longer, investigative pieces. Instead, the goal is to provide more context to issues facing the Latinx community: "I want to just say that we're nobody's first news source. So, over the years, we've seen our mission more as about context and analysis. We'll not often do a show about a breaking news event."

This means finding the balance between seeking out stories that are current with those that are evergreen. At times, the team will seek out stories that are often ignored by large, commercial news outlets. Or they try to spend more time on a current issue, but give a deeper, more thoughtful analysis. According to Bishop:

> We'll do [a segment] three months later and really try to look at root causes and history and looking at sides of history that are generally not told, is a huge part of what we do. Whether it's talking about the history of Latinos in Rock and Roll, which began with a question about why the guitar even exists in the United States of America. It's because of the vaqueros. . . . These are the kinds of stories that don't see the light of day because of the way that history is written.

According to Reiman, the team at *Latino USA* holds regular planning meetings, which are meant to chart out longer-term production plans. At the same time, they try to be responsive to what is happening at the time. For example, during the pandemic of 2020, *Latino USA* produced stories about the COVID-19 pandemic's impact on Latinx small businesses (Hinojosa 2020d) and the health risk to Latinxs who are in ICE detention centers (Hinojosa 2020e).

Reiman described it as a collective process in which the team will vote for stories they would like to pursue. If there's enough support, Reiman will defer to the group despite any reservations she might have. She told me that the team at *Latino USA* also looks for stories that are compelling but have strong narrative potential, and she particularly likes pitches for stories that she feels audiences have not heard before.

Reiman characterized *Latino USA* as a "storytelling space," one with good narrative journalism. She likes stories that are character driven and sees a distinction between a good idea or subject versus a good radio story. As a producer of an audio story, sound is also important to the process. Reiman thinks of her work as making an audio film, but, instead of being visually focused, she must attend to language, music, ambient sound, and other sonic considerations.

At the same time, Reiman believes that *Latino USA* has a mission to inform the public. This might be accomplished by integrating the perspectives of sources who can provide context, what she terms *explainers*. For example, the team likes to bring on experts who can provide alternative points of view, which, she believes, provides a deeper understanding of a particular issue. According to Reiman, "it's like the *New Yorker* of the Latino community."

Latino USA and the Ideal Listener

Reiman's description of *Latino USA* as the "*New Yorker* of the Latino community" raises the important issue of audience. When Harold Ross co-founded the *New Yorker* in 1925, it was unabashedly directed toward a highly selective reader, one who was educated and urbane (Lepore 2010). In many ways, the listener for *Latino USA* tracks with NPR's larger audience of educated, middle-class listeners. According to NPM (2017), 73 percent of *Latino USA*'s listeners are college educated, and over half have an average annual household income of at least $75,000.

Podcasting has enabled *Latino USA* to expand its distribution, but it has not significantly diversified its audience. According to *Latino USA*, 88 percent of its listeners are college graduates, and almost all have an average annual household income of at least $75,000. Of all *Latino USA*'s podcast listeners, 55 percent are thirty-four years old or younger. Until Hinojosa parted ways with NPR in 2020, *Latino USA* was the network's top program for Latino listeners, but this speaks more to the network's overall lack of Latinxs. Only 27 percent of *Latino USA*'s podcast listeners are Latinx (National Public Media 2017).

The team is not unaware of the dilemma of reaching a primarily white audience with Latinx-oriented content. "Yeah, but it's public radio," one staffer told me, "and public radio is a very white space." This is certainly true, but, as a program that purports to speak on behalf of the Latinx community, the nature of the audience raises some questions about the actual mission of *Latino USA*. Hinojosa described her goal to educate NPR's audiences about the Latinx community while continuing to grow its Latinx audience:

The truth is, and you know this, the majority of our listeners on terrestrial radio are not Latino. So, we have this very particular responsibility to be able to talk and report for Latino listeners, so they understand that we totally get it, but also not make our non-Latino listeners feel like, "I totally don't get this." And there's a magic there that this team knows how to do. So, we have to navigate being really inside baseball, and also being in the bleachers at the same time.

Here, Hinojosa is describing the editorial decisions needed to ensure that the program will speak to multiple publics. In previous interviews, Hinojosa has spoken about the challenge of producing a magazine show that must speak to a broad audience. For example, she spoke to *Urban Latino* magazine's managing editor, Juleyka Lantigua (Nieman Reports 2001):

There is no set audience for our program. To us, it is the broadest possible audience. We know that we're not only speaking to Latinos. Therefore, I don't keep anything particular in mind when reporting for *Latino USA*. Perhaps we use more Spanish. That's purposeful, so that we engage listeners. We want them to hear how we speak. We work on an assumption that the listeners of *Latino USA* are knowledgeable; that we don't necessarily have to explain who Gabriel García Márquez is. We don't have to explain what an undocumented immigrant is. We assume people already understand such things and that it's part of why they listen to *Latino USA*.

As Hinojosa notes, part of this process involves making specific choices about which cultural references to include, and the degree to which Spanish should be used. Many of the journalists and producers at *Latino USA* are proficient, to varying degrees, in Spanish and English. In this case, the practitioners at *Latino USA* have linguistic capital, meaning that their ability to speak in multiple linguistic codes allows them to navigate this tricky landscape. Many of the subjects they are interviewing are either Spanish-dominant or bilingual. Given the permissible alternatives, this means that *Latino USA*'s practitioners are continuously making strategic choices about what is the best alternative, given the conditions of the interview.

When I spoke with the production team at *Latino USA*, they indicated that they are trying to remain faithful to the linguistic practices of the interviewee. "If the character that we have code-switches," one staffer said, "we leave that." They also want to remain faithful to their own way of speaking. "We think about how we talk," one journalist reported. "I mean, we want our stories to sound conversational and natural. If we were to code-switch ourselves, we would write it that way." Conversely, the team indicated that they do not want to sound disingenuous by inserting Spanish gratuitously. "I'm not a big fan of writing pointless Spanish. We have a style guide, and we say don't call a stomach a *pansita* without good reason."

However, the *Latino USA* staff was also aware of the practical considerations involved in creating a radio show. *Latino USA*'s producers face a number of implicit and explicit restrictions, which were shared by other public radio practitioners with whom I spoke. First, they recognized that they did not have unlimited time to tell a story, which means that they are attentive to the need for efficient communications. Second, *Latino USA* is an English-language program that reaches primarily English monolinguals, and they recognize that they must be attentive to the needs of their audience. "If it happens organically, we won't edit it out," Reiman said, "but we're never going to leave untranslated giant chunks of Spanish because that's not what we want to do to the non-Spanish speaking listener." Striking the right balance means looking at any given segment holistically. If the producers believe there are enough contextual clues, then they will include Spanish.

The Business of *Latino USA*

When I asked the team how they came by their stories, Hinojosa told me that "it's a nice mix of on the ground reporting, [research] we do here, us traveling. Freelance collaborations." In other words, a given story might come to *Latino USA* in a number of ways. Some stories are pitched by freelance journalists. Others are a direct response to the news cycles. For example, "Digging into 'American Dirt'" was produced expeditiously in an effort to capitalize on the interest that had been gaining momentum on social media. Other stories are the result of research conducted by the team at *Latino USA*. "I do a lot on Reddit," one staffer told me. "And if it's something interesting, I reach out to people."

Hinojosa is central to this process, and she often becomes a conduit for potential stories. Hinojosa travels extensively across the country for speaking engagements, which gives her a sense of what is occurring nationwide. "I'm experiential," she told me, "so whatever I'm feeling out there." On occasion, producers and journalists have used these trips as a way to gain some preliminary insights into potential stories. "We use that," one the team member told me. "If María is going to be traveling to Nebraska, then we might dig and see what we might do."

The Strange Death of José de Jesús is a good example of this process. According to Bishop, one of the producers of the story, the idea for that story began with a business trip that Hinojosa took to Arizona. "We're pretty sure that María had an event in Arizona and we thought about how to leverage it." Bishop reported that a story based in Arizona was appealing because Arizona is an important market for *Latino USA*. Once he knew that Hinojosa would be traveling there, Bishop got on the phone with some of their local contacts: "I would often try to call different organizations and get tips. In this case I had called Puente, which is an activist organization based in Phoenix. And I asked them, what are the stories? What's happening now? What should we know about that's going on? And they told us about this unexplained death that they thought was fishy."

Once he had gathered the preliminary background information, Bishop then pitched the story to Hinojosa. Together they discussed its potential: "And so, we started researching that. We thought it was really interesting. María, all credit to her, saw that there was something there. This drama, this man who choked on his own sock. It was just horrifying. And it was a starting point. And I think, they [Puente] were so sure that there was foul play involved. And that was part of our interest in the story. Could there have been?"

Once it became clear that this could be a compelling story, there were discussions about the kinds of resources the organization should invest in its production. In this case, it appears as if the decision to pursue the story was part of a larger creative and strategic discussion. The team decided that the story required a much longer format than what they typically produce. They also saw this as an opportunity to engage in investigative journalism, which requires the time to follow up on leads and the resources to travel. "We spent a lot of time and a lot of money on this [story]," Bishop told me. "We took a lot of trips. We went to Mexico. We went to Arizona three or four times."

As Bishop indicates, it takes money to do this kind of journalism. Some of the funding for *The Strange Death of José de Jesús* was made possible through a grant from the Marshall Project, a nonprofit organization designed to raise awareness of issues in the criminal justice system. However, *Latino USA* must be selective about how it uses its limited resources. Compared to larger, better-funded national news organizations, *Latino USA* is a modest operation. "It 100% affects everything," Reiman told me about funding. "We are small, but ambitious. And sometimes reality gets in the way. At the end of the day, we produce one hour of radio content a week and we have eight people to do it. We have to hire a lot of freelancers." According to Bishop, this means that the staff at *Latino USA* must be entrepreneurial in their approach: "Sometimes there's a little bit of chaos in terms of funding. All of a sudden, money's dried up. It hasn't been my job luckily, but we'd have to pull up our bootstraps and figure out how to do everything with less."

Latino USA's funding primarily comes from foundations, which accounted for 86 percent of its revenue in 2017. Individual supporters provided the second-largest amount, accounting for 8 percent of the show's revenue (Futuro Media Group 2017). The sponsors that support the show include a mix of private and philanthropic organizations that include the Ford Foundation, The Annie E. Casey Foundation, the CPB, and Carnegie Corporation of New York, as well as private companies such as Deutsche Bank and the Toyota Foundation.

Reiman credited Futuro's development team for finding the resources for *Latino USA* to fulfill its mission, but she also credited Hinojosa for continuously promoting the program. Hinojosa has been a tireless advocate for the brand, and she is highly active on various social media platforms as a way to promote upcoming podcasts. However, during her relationship with NPR, there were limitations to what Hinojosa could control. As a primarily terrestrial show, the success of *Latino USA* was dependent upon its distribution. The greater the number of NPR stations that carry *Latino USA*, the larger the audience that it can deliver. The larger the audience, the more appealing *Latino USA* becomes to prospective sponsors.

However, station managers and programming directors serve as important gatekeepers in this process. NPR member stations have the option to pick up *Latino USA*, but NPR cannot force stations to carry the program, and it can be a challenge to get station managers to see the program's value. Being a Latinx-oriented news program raises unique considerations for station managers. First, station managers may choose to air *Latino USA* because they believe the program may appeal to audiences in their communities, particularly in markets with high numbers of Latinxs. Second, it may signal to key stakeholders that the station is committed to diversity programming. As one station manager told me, "absent any diversity in station, CPB serving stations are like, how are we doing on diversity? Oh, we have *Latino USA*. And before the program died, *Tell Me More*. So, we can say we are plenty diverse, but it's bullshit. It's just token diversity."

As the station manager suggests, the show can be subject to racializing practices. Hinojosa and her team believe that some programming directors have biases that restrict their ability to see value in a Latinx-oriented show: "The problem is that we're dealing with the structural, frankly, the reality of a system that was built on the commitment to diversity, as Bill Siemering said. And I was lucky enough to know Bill Siemering. But to actually make that happen, it's been a little bit of a challenge."

According to Hinojosa, NPR's commitment to diversity has never been fully actualized. She believes that station managers are generally conservative in their thinking. "If the programming directors are seventy-year-old white guys," Hinojosa asked, "what's their impetus for changing things or wanting to listen? And it's a real problem for us." Hinojosa does believe that some progress has been made, but is apprehensive. "At the CPB, they're trying to create more diversity," she told me. "Yes, it's happening, but we've been talking about this for twenty-five years."

I found that there are a number of reasons why a station manager may not choose to air *Latino USA*. Ultimately, the number of programs that any member station can build into its weekly schedule is finite. Therefore, station managers must make choices about which programs they believe will deliver the most desirable listeners: those who are likely to support their stations financially. Others have struggled to find a way to seamlessly integrate the program into their weekly schedule. For example, when I asked one station manager why he has decided to include *Latino USA* in his station's weekly rotation, he replied "honesty, I inherited it." The participant also reported that he had inherited another Latinx-oriented show that was locally produced and focused on the state's Latinx community. According to the participant, both shows are included, but not well integrated: "There's almost no interaction between [the local program] and our program director. So it just sort of flounders on Sunday night. And so *Latino USA* is adjacent to it, and that's kind of like the Latino barrio."

The station manager went on to explain that a single show focused on Latinxs is only a stopgap solution to the larger issue of diversifying the newsroom. But when I asked him how he addresses this larger issue, he acknowledged that he has neither the staff nor the resources to cultivate journalism that will serve the growing number of Latinxs in his market. Like other member stations, time and money are limited, which means there is little opportunity to hire Latinx talent or spend time building inroads into the Latinx community. Under these conditions, *Latino USA* becomes the only form of Latinx-oriented programming on his schedule.

Latino USA in the Time of Political Disruption

When I spoke with the team at *Latino USA*, it became evident that they have a strong sense of mission. In its public-facing documents, Futuro embraces a traditional understanding of the press, which is, in their words, to "serve as watchdogs over public affairs and news access." This vision is built into Futuro's mission statement, which reads: "The journalism of the Futuro Media Group will hold those in power accountable and support the open and civil exchange of views, even views that staff disagree with or may find repugnant."

In practice, however, *Latino USA* is consistent with NPR's overall approach to journalism, which is to inform the listening public, but not necessarily to engage them civically. I asked the team how they thought this kind of civic engagement compared to stations that have taken a much more active role in engaging Latinx listeners. For Reiman, there is a distinction between civic engagement and what she describes as "having a more informed, civically engaged society." Reiman's goal is to provide a more thoughtful, nuanced story about policy issues. Listeners may or may not choose to act on that information. According to Reiman, her goal is "to present it in a way that makes you think. And that may affect how you vote or participate in your community."

The team believes that they can serve an important role simply by telling the stories of Latinxs in their own voices. "We're responding to an invisibility," Hinojosa told me. "They [Latinxs] still feel invisible. We're starting off from the point of responding to that invisibility." However, the team acknowledged that the responsibility to report on Latinx issues had escalated with the election of Trump. "We have our work cut out for us," Hinojosa said. "The narrative about Latinos/ Latinas in this country, that we had nothing to do with creating, has dehumanized us in many ways." To respond to these public discourses, the group feels some obligation to tell more-nuanced stories that humanize Latinos.

Additionally, the team felt a particular responsibility to report on immigration, which had become the centerpiece of Trump's domestic policy. But the severity with which the Trump administration had pursued this policy raised a number of human-rights concerns, which created urgency in their work. According to Hinojosa:

> I do think we also feel a tremendous amount of responsibility to report on what's happening with immigrants, while we also feel the reality of "we're tired of telling this story." This is not the only story about Latinos, Latinas. And yet at the same time, the historical dynamic in this moment which is pushing us to absolutely tell that story. We are the eyes and ears and we must absolutely tell that story.

Latino USA's focus on immigration has subjected Hinojosa and her team to accusations of bias in ways that other journalists are not. This is not necessarily surprising. As Marchetti (2005) argues, journalists who are considered "specialists" are often stigmatized as having been captured by their sources, or even as serving as de facto spokespersons for the communities they cover. In the same way, Hinojosa and her team have been accused of being advocates for the Latinx community. During our meeting, Hinojosa made sure to distance herself from the term "advocate." According to Hinojosa:

> We're not advocates. That's not the way we start. You will never hear that term come up in a news meeting. Like, "how about if we use this piece to advocate for. . . ." No one would ever say that because we're journalists. I guess if we were asked, what do you advocate for? It's probably like the constitution, freedom of expression. Equality under the law. That's the kind of stuff we would advocate for.

While Hinojosa believes that journalistic standards must be followed, she has not been shy about challenging normative practices within journalism that she believes to be harmful. Hinojosa has been openly critical of newsrooms that purport to be objective but that, in fact, promote a particular ideological point of view. She is particularly attentive to language and the ways in which it can frame particular communities. Hinojosa told me: "Illegal immigrant. We don't use the word illegal to define someone. They could be living here illegally or cross here illegally. The other one that we try to avoid as much as possible is the word 'minority.'"

In her interviews and speaking engagements, Hinojosa has consistently made the case that *illegal* is problematic, because it has become the dominant frame through which we see Latinxs, thereby impeding our ability to see their full humanity. Consider an article written for *Change Agent* (2019f), where Hinojosa tells the story of Estrella, a trans woman who was detained by immigration officials after she had reported domestic abuse. According to Hinojosa, journalists fail to adequately tell the story of someone like Estrella: "Local news covered Estrella's story. So did *The New Yorker*. But it wasn't until journalist Jonathan Hirsch and I interviewed her in prison that we heard her voice. Her own voice. Telling her own story."

Here, Hinojosa describes the double bind that someone like Estrella faces. She is Spanish-speaking, which means that she must rely on English monolinguals to faithfully tell her story. Second, the designation of *illegal* foregrounds her citizenship status, while background-ing other more important facets of her identity. Hinojosa (2019f) con-tinues: "If we only used terms the U.S. government chooses for people like Estrella, we'd be dishonest. Estrella is not an 'illegal.' As I've said many times before, 'illegal' is not a noun. Estrella is not 'sexually con-fused.' Estrella has identified as a woman as early as her teens. If we use language that misnames, we cannot tell her story properly."

By describing this process as *misnaming*, Hinojosa is invoking Valdivia's (2010) point that naming is an intensely political act—and an exercise of power. To Hinojosa, the term *illegal* can obscure, rather than reveal, thereby undermining the very point of journalism. By choosing to use certain words over others, news organizations wield a tremendous amount of power to frame reality. However, Hinojosa's decision to avoid the use of *illegal* when possible is not quite consis-tent with NPR's own editorial standards. While NPR has acknowl-edged the problematic nature of the term, they do not have a formal policy that bans the use of the word *illegal* (Memmot 2017).

Latino USA in a Changing Media Landscape

It is important to think of *Latino USA* as an organization that is evolv-ing within a larger media landscape that is itself evolving. The number of Latinxs has grown, but so too have other nonwhite groups, prompt-ing Futuro to define its overall audience more broadly. According to their website, Futuro's mission is to "create multimedia content for and about the new American mainstream in the service of empower-ing people to navigate the complexities of an increasingly diverse and connected world" (Futuro Media Group 2020).

Two of Futuro's products, *In the Thick* and *America by the Num-bers*, focus on people of color generally, while two others are Latinx specific. In 2018, Futuro acquired *Latino Rebels*, a digital news outlet founded by Julio Ricardo Varela. Then there is *Latino USA*, which re-mains the centerpiece of Futuro's product portfolio. As Bishop stated, "it's still the tentpole. By far, the main thing that we do. It has the largest staffing component. And we're thinking that it all interacts with each other."

The kind of growth *Latino USA* has enjoyed in recent years has been made possible, in part, by the digital and political disruptions that have transformed the field. Several of Futuro's properties are primarily digital spaces. Furthermore, distribution through online platforms has also allowed *Latino USA* to rely less heavily on station managers, who have traditionally served as gatekeepers. According to Bishop: "I would say that every year terrestrial radio is becoming a smaller piece of the audio pie, in terms of listening, in terms of relevance, in terms of a lot of things. I think it's really so fundamental and can never be neglected because the radio really gives us access to many ears and minds to our work that would never look for us and find us, necessarily."

Bishop describes podcasting as slowly changing the dynamics of public radio, but he also feels there is a way to go: "Podcasts have become kind of this elite product in a lot of ways. I think it's changing slowly as more people are coming to the table, but for a long time, the audience was overwhelmingly college-educated, wealthy, and white. And that's true across public radio, which is obviously a huge problem. And a problem with the very concept of public radio as it was envisioned."

Here, Bishop brings up the unique role of public radio in the current landscape. To get a better sense of how the team envisions the unique mission of public radio, I asked them how they measured the success of their work. In some cases, it was based on internal markers. "Recognition from our peers, from people that value good work," one staffer told me. "If we all liked it," another told me. This attentiveness to internal validation is consistent with Bourdieu's (1993) point that producers seek cultural legitimacy, which comes not only from a public, but from a public of equals who are also competitors.

But Bourdieu distinguishes between autonomous fields and fields that are heteronomous, meaning that they are subject to external pressures, such as market forces. Because *Latino USA* is dependent upon listeners and outside organizations for funding and distribution, external measures were equally as important. The team indicated that they pay attention to increases in listener downloads. It is also important when the program gets picked up by a member station in a large media market. These figures are featured in public-facing documents. The team also mentioned that they are attentive to the feedback that they receive online.

Industry awards are another way of validating their work. Over its lifetime, *Latino USA* has received a number of awards, which are touted on Futuro's annual report and their website. "We've won every major award except the DuPont," the team proudly reported. "We need the Livingston and then a Pulitzer." I asked Bishop why industry awards are important to the organization and its employees, to which he responded:

> It's a way of celebrating us. It's a hard job. A lot of peo-
> ple in journalism, especially now, are disparaged all the
> time. It celebrates our accomplishments. You can make
> a criticism that they're self-aggrandizing, but I think for a
> program like ours, it's important. I've felt like Futuro has
> had to validate itself in the larger eco-system. For a lot of
> reasons. For one, we're small. And there's an implicit bias
> against a Latino organization, fundamentally, a Latino
> organization, and what we're doing. And so, winning the
> Peabody, winning RFK's. It shows that even though we
> don't have the resources, we can do work that matters.

As Bishop suggests, industry awards can serve as a form of objectified capital, which can then be converted into economic and social capital. For example, awards can serve as validation for foundations that their money has been well spent. It can also be instrumental in recruitment of new talent and can translate into new kinds of partnerships.

Given their reputation for producing quality content about the Latinx community, *Latino USA* has been able to develop new relationships with both commercial and noncommercial media organizations. In addition to its work with NPR and PBS, the organization has also developed content in partnership with NBC News (2016) and the *Los Angeles Times* (2019). The kinds of partnerships in which Futuro is engaging is reflective of a landscape in which boundaries between commercial and public media are becoming more fluid.

At the same time, the demand for this kind of content is, in some ways, a product of the political turmoil that existed during the Trump presidency. As Trump has made immigration the centerpiece of his political agenda, news organizations responded by dedicating more resources to reporting on the topic. Because the program arrived in the market early, *Latino USA* has essentially established itself as the brand leader in Latinx-oriented public radio content, billing itself as the "longest-running Latino-focused program on U.S. public media" (Futuro Media Group 2020). But, according to Reiman, more established players are entering the field, creating more competition within the field: "We were the ones that were telling this story before it was sexy to do so . . . this show has understood the nuance of what it means to be an immigrant, the complexity of the border, the complexity, the nuance, which now every journalist is parachuting in and doing a very sexy story about this."

These new players are now competing for a limited number of resources, which includes listeners, funding, and talent. Soon after participating in this interview, for example, Reiman left *Latino USA* to work as a producer for *This American Life*, one of the more dominant players in podcasting. There, Reiman was part of a team that won the Pulitzer Prize for a story called "The Out Crowd," which focused on Trump's "Remain in Mexico" policy (Glass 2020).

The ability of *This American Life* to win the industry's most coveted prize by covering immigration issues, while siphoning off talent from *Latino USA* in the process, must be seen as a threat to the organization, in which there is fierce competition for limited resources within the field. Bishop, however, has a different perspective.

> I think it's interesting for *Latino USA* to see, more broadly, the issue of immigration covered with the resources that it has been, the past four years by the entire media eco-system. And for us, trying to figure out, well, we're not *The New York Times*. We're not NPR. We're not any of these. We're a tiny non-profit, with a lot of heart and some good contacts. But how do you tell these stories? I mean we're not going to break immigration news before any of these people. I think there's been some frustration seeing the way that us, being there first, and doing a very serious job of covering Latino issues for so long, and then seeing so many other people start to do it.

Yet Bishop feels optimistic. From his perspective, the entire market for Latinx-oriented news content is growing, which is not only a boon for the professionals who produce that content, but for also for

Latinx listeners. "I think ultimately it's good for everyone," he told me. "It's good for the country. It's good for Latinos."

Despite increased competition for resources, the emergence of new players within the field has also opened up new opportunities for Futuro. In 2020, Hinojosa made the remarkable decision to end her long-standing partnership with NPR, which had managed the program's national distribution since 1994 (*Latino USA* 2020). Instead, Hinojosa opted to partner with Public Radio Exchange (PRX), a nonprofit media company that distributes other successful public radio programs, including *The World, This American Life,* and *The Takeaway.* The new partnership promises to expand Latino USA's reach. According to PRX, the company reaches more than 28.5 million listeners and generates over 70 million podcast downloads per month (PRX 2020).

During an interview with NPR's *It's Been a Minute with Sam Sanders* (Sanders 2020), Hinojosa was asked about her decision to part ways with NPR. At first, Hinojosa framed it primarily as a business decision: "NPR's my family, *mi familia.* It was my first job. But I started a company and I have a team of very savvy business and media executives and journalists. And when they said, look, we have an opportunity here in a competitive market. PRX, somebody who really wants to go big."

But Sanders pressed on, asking Hinojosa whether she believed that NPR neglected the show or simply did not promote it enough. Hinojosa was reflective: "*Latino USA* right now is growing an audience at extraordinary numbers. I think we're one of the few public radio programs that was previously distributed by NPR that is growing audience with these numbers. And so the fact that we made this decision says everything about what NPR thinks about *Latino USA.*"

Hinojosa went on to admit that she believes that NPR simply did not support a Latinx-oriented show. "There was a point not too long ago when one of your colleagues called me up," she told Sanders. "She's a Latina colleague at NPR in the newsroom, and she called me up and said, 'do you think *Latino USA* has been this incredibly successful because of NPR or despite NPR?' And no one had asked me that. And I kind of went, hmm. And I said, well, actually despite."

This revelation comes as no surprise to María Martin, who has been openly critical of NPR over the years. "My hopes for greater investment by NPR in *Latino USA* were misplaced—for myself but more for the audience that radio was meant to serve," Martin wrote in her memoir, *Crossing Borders, Building Bridges* (2020, 59). "Throughout the eleven years I produced the program, and even after that, I believe the network largely treated the project as a token step-child and did not give it the support it deserved."

Figure 1 María Martin consulted during the initial development of the show. She later served as *Latino USA*'s executive producer. Courtesy of Nettie Lee Benson Latin American Collection, University of Texas Libraries, The University of Texas at Austin.

Figure 2 María Hinojosa was hired as host of *Latino USA*. In 2010, she founded Futuro Media Group, an independent nonprofit media organization, which continues to produce the program. Courtesy of Nettie Lee Benson Latin American Collection, University of Texas Libraries, The University of Texas at Austin.

MAY 1 8 1993

UT News

THE UNIVERSITY OF TEXAS AT AUSTIN

News and Information Service
P.O. Box Z, Austin, Texas 78713-7509
(512) 471-3151 FAX (512) 471-5812

Contact: Becky Knapp
Date: May 14, 1993

Latino USA Takes to the Air

AUSTIN, Texas -- For the Latino community, it's the biggest crossover endeavor to hit the air waves since *La Bamba* took the country by storm five years ago.

It's *Latino USA*, a weekly journal of news and culture, produced in English, that promises to heighten national awareness on the Latino community. The 30-minute program, which airs on Fridays at 4:30 p.m., is produced by the Center for Mexican American Studies at KUT 90.5 FM, on the campus of the University of Texas at Austin. The series is currently distributed, through the Longhorn Radio Network, to more than 70 non-satellite stations from Maine to Washington.

On July 2, the show will begin satellite broadcasting, opening the program to some 460 other stations across the country.

"I think this program fills a need in public radio," said Dr. Gilberto Cardenas, director of the Center for Mexican American Studies and executive producer of *Latino USA*. "There is not a single program except ours that serves that entire Latino community -- Mexican-American, Cuban, Puerto Rican, and Central and South American. And because the program is in English, it will enable us to broaden the audience. The message we want to convey is that this program will allow the Latino communities to learn more about each other, and will also allow other people throughout the nation to learn more about Latino communities."

There are many multicultural radio programs available now, Cardenas said, "and we are not looking to replace those, because they serve an important function. But they don't provide national distribution, in English, of a program that targets a Latino audience, and we feel we can fill a much-needed gap there."

Latino USA programming will basically fall into five major categories: topical interviews, panel discussions, youth issues, biographies, and commentary on such issues as social policies, the arts, culture and entertainment.

In addition to Cardenas, the staff at *Latino USA* includes Vidal Guzman, marketing manager and social producer; award-winning producer Maria Martin, former national news editor with National Public Radio; New York-based program hostess Maria Hinojosa, Elaine Salazar, marketing and fundraising, and Dolores Garcia, marketing assistant.

A number of interns, work-study students and graduate and undergraduate production students round out the staff in research, editorial and production roles.

(over)

Figure 3 A press release from the University of Texas press announcing the show and explaining *Latino USA*'s basic premise. Courtesy of Nettie Lee Benson Latin American Collection, University of Texas Libraries, The University of Texas at Austin.

Figure 4 Novelist Daniel Alarcón co-founded *Radio Ambulante* in 2011. Photo by May-Li Khoe. Image courtesy of Radio Ambulante Studios, Inc.

Figure 5 Co-founder Carolina Guerrero serves as CEO of the organization. Image courtesy of Factual and Radio Ambulante Studios, Inc.

Figure 6 Carolina Guerrero and Daniel Alarcón host a fundraiser in Oakland in 2012. With funds raised from a Kickstarter campaign, the fledgling podcast went on to become a global enterprise. Image courtesy of Carolina Guerrero.

Figure 7 Felix Contreras was a producer and reporter at NPR's arts desk before co-founding *Alt.Latino*. Photo taken by Allison Shelley. Courtesy of NPR.

Figure 8 Co-founder Jasmine Garsd was a co-host of *Alt.Latino* before leaving NPR in 2016 to work on the PRI/BBC project Across Women's Lives. Garsd returned to NPR as a national correspondent in 2021. Photo taken by Yanina Manolova. Courtesy of NPR.

Chapter 4

RADIO AMBULANTE

Entonces, asumamos que, por la misma naturaleza de nuestro podcast, nos interesan las fronteras. Por todas las razones que se pueden imaginar. Son lugares de tránsito. De fluidez cultural y lingüística. Son zonas de intercambio. Depende de dónde las mires, claro, pero a veces esas mismas fronteras hasta pasan desapercibidas.

— *RADIO AMBULANTE*, UNA CIUDAD EN DOS

So, let's assume that because of the very nature of our podcast, we're interested in borders. For all the reasons you can imagine. They're places of transition. Of cultural and linguistic fluidity. They're places of exchange. It depends on where you're looking from, of course, but sometimes the borders themselves go unnoticed.

— *RADIO AMBULANTE*, A CITY IN TWO

In his book, *The Idea of Latin America* (2005), Walter Mignolo argues that maps are, in essence, fictions. While maps purport to represent a world that is "objectively there," they are better understood as social constructions that advance inherently ideological positions. Mignolo then considers the process by which the idea of a single America began to be divided according to its imperial history, placing Anglo America in the north and Latin America in the south. In this new configuration of the west, Latin America became a sort of marginalized racial category, based loosely on geographic location, skin color, and language. This was seen in opposition to a racially pure, English-speaking North America. Mignolo further argues that the growing presence of Latinxs has interrupted the fantasy of the United States as a white space, which has become a source of anxiety for Anglos, who fear they are losing their identity as a nation.

Some of these themes were explored in an episode of the Spanish-language podcast *Radio Ambulante*. Titled "Una Ciudad en Dos" (Alarcón 2020), the episode focused on the unique relationship between El Paso, Texas and Ciudad Juárez, Mexico. For almost two hundred years, these cities existed as one, before the Mexican-American War divided the region in two. For centuries, this region has been linked across a geopolitical boundary by economic, social, and cultural exchange. However, in recent years, the physical and symbolic boundaries that separate the two nations have become more pronounced and policed.

These tensions boiled over when a gunman had opened fire at a Walmart in El Paso, killing twenty-three people and injuring twenty-six others. In a manifesto released before the shooting, the gunman began to fixate on the threat to the region's racial purity, speaking of a "Hispanic invasion of Texas" (Brockell 2019), a claim that builds on discourses in which the original inhabitants of the region (Native Americans and, later, Mexicans) are recast as "invaders" against whom the territory must be defended, often through violence (Luiselli 2019).

But the specter of a Latinx invader infiltrating a white Texas obscures reality of life on the border, and the podcast explores this reality deftly. Borderlands are spaces of hybridity, where Mexican and American citizens meaningfully interact, and where English or Spanish is widely spoken. The arbitrary nature of borders is addressed explicitly by journalist Victoria Estrada, who guides the listener through the episode:

> Y es que es eso lo que los mapas no muestran: que la frontera no es una línea, sino una región, con una cultura, un pueblo, un lenguaje particular. El biculturalismo es una cultura en sí misma, igual que el binacionalismo es una forma particular del nacionalismo.

> And that's what the maps don't show: that the border isn't a line, but a region, with its own culture, people, language. Bi-culturalism is a culture in itself, just like bi-nationalism is its own form of nationalism.

In a number of ways, "Una Ciudad en Dos" reflects the very ethos of *Radio Ambulante* itself. During a visit to Notre Dame University (Moreno and Anderson 2014), Daniel Alarcón, host and co-founder of *Radio Ambulante*, spoke about his vision for the podcast, which, at the time, was in its infancy:

And it is certainly something we have been thinking about . . . with *Radio Ambulante,* our radio project, that is thinking about the Americas as one cultural region. In our case we are thinking of it as a cultural region united by Spanish, a region that would include Spanish-speaking Latin America but also the United States.

This perspective distinguishes *Radio Ambulante* from NPR overall, which is grounded more firmly in the idea of the nation-state. And while NPR has seventeen international bureaus, they operate more like satellites that serve at the behest of a centralized operation based in Washington, D.C. Only two of these bureaus are in Latin America.

By contrast, *Radio Ambulante* embraces the idea of a post-national world, in which borders are increasingly irrelevant. Furthermore, the producers of the show explicitly embrace an all-encompassing vision of America, and the stories that are produced by *Radio Ambulante* take place in countries throughout North, Central, and South America, as well as Spain. *Radio Ambulante*'s transnational perspective also distinguishes it from *Latino USA,* which was first conceived of as a show for Mexican Americans, then later for U.S. Latinxs generally. The two shows also differ in their relationship to technology. *Latino USA* began as a terrestrial program, which later expanded into digital distribution. By contrast, *Radio Ambulante* was conceived of as an entirely digital enterprise. Consequently, its producers have benefitted from the transnational flows and online networks that have replaced traditional geographically fixed spaces as primary forms of identity. The growing irrelevance of physical space has allowed for a relatively small organization to develop a transnational professional community, to build collaborations with journalists based throughout the Americas, and to reach listeners beyond the boundaries of the United States.

Audio-Storytelling in an Age of Digital Disruption

Radio Ambulante is a long-form narrative audio-storytelling program that is produced in Spanish. The episodes that *Radio Ambulante* produces range in length from a half-hour to an hour, but each program is structured with an introduction, climax, and denouement; has its clearly defined main characters; establishes the setting; and provides enough context to enable listeners to get the overall narrative. Almost all of the podcasts are built around interviews conducted by journalists (Fernández-Sande 2014).

Radio Ambulante was co-founded by Carolina Guerrero, who serves as the project's executive director, and her husband and business partner, Daniel Alarcón, who serves as the program's host. When *Radio Ambulante* was first conceived of in 2011, public radio was undergoing a significant transformation with the emergence of podcasting.

When Dave Winer coined the term *podcasting* in 2004, the medium itself was an emerging phenomenon. The *pod* in podcasts stood for *personal option digital* (or *personal on demand*), and they were relatively easy and inexpensive to produce (Walker 2019). Developments in communications technologies have made it possible for individual producers to create content without the support of stations or radio networks, leading to a surge in creativity and diversity within the space. Some of these podcasts were able to serve the needs of the public without getting any direct financing from the CPB.

This came at a time when NPR was struggling to reach younger listeners. According to Falk (2015d), the terrestrial audience for NPR has been declining, but there has been growth in NPR's digital platforms. Furthermore, because of the relatively low cost, NPR could support a program intended for a niche, rather than broad-based, audience. However, NPR's podcast strategy has been an evolving process. According to Eric Nuzum, NPR's director of programming and acquisitions, there was little strategic direction when NPR entered the space. First, there were more than one hundred different podcasts, making it difficult for consumers to find any given program (Mullin 2015).

So, NPR decided to cull the number of podcasts it offered, eliminating shows that did not meet NPR's standards. As Nuzum stated, "podcasts that consisted primarily of excerpts from other shows were eliminated." NPR also cut a number of shows that were essentially roundups of stories about movies, science, or international news. With a few exceptions, "anything that wasn't a 'full experience.'" Nuzum went on: "a standalone podcast that didn't need to borrow from other NPR offerings—was cut. The next step in NPR's strategy was to bring a wider audience to this leaner list of shows" (Mullin 2015).

Today, NPR promotes a number of podcasts that are organized by topics such as science, arts and culture, business, and politics. Until 2020, when *Latino USA* decided to end their distribution partnership with NPR, the network promoted only three podcasts that were Latinx centric, but it is important to understand their relationship to one another. *Latino USA* is a news-magazine format, produced in English and focused primarily on the U.S. Latinx community. *Alt.Latino*, on the other hand, is a music-oriented podcast, which is also produced primarily in English, but has a transnational sensibility and often features artists who are not American. And then there's *Radio Ambulante*, which, according to NPR's podcasting director, is "a Spanish language podcast that uses long-form audio journalism to tell neglected and under-reported Latin American and Latino stories" (National Public Radio 2020c).

Ideally, NPR can maximize synergies among all three programs while avoiding too much overlap in content or audience. Generally, each of these programs operates somewhat independently, but when I spoke with Marlon Bishop, producer at *Latino USA*, he mentioned that there have been moments in which they covered the same topics. "There have been funny times when we covered the same stories. We've collaborated as well. Specifically, Peru and the World Cup was covered by *Ambulante* and us, weeks apart. But they were very different stories with different audiences."

As Bishop points out, despite the overlap, *Latino USA* and *Radio Ambulante* conceive of their audiences differently and deliver different kinds of stories. Over the years, the producers of *Radio Ambulante* have attempted to clarify its brand positioning within a crowded marketplace. During an interview with Nieman Reports (2020), Alarcón framed the program this way:

> We're not the newscast on NPR. We are much more like a long-form piece you might read in *The New Yorker* or a story you might hear on *This American Life*. We've done stories investigating the punk rock scene in Cuba, the lost orphans of a landslide in Colombia, and the students in Chile who took over a university building and burned all the receipts of their debt and what happened as a result.

Certainly, these kinds of stories are unique for NPR, which has traditionally overlooked Latin America. This ability to bring a different focus to a large and heavy organization like NPR is remarkable. It is equally remarkable that they have been able to do this in a relatively short amount of time and with little to no radio production experience. Their ability demonstrates the ways in which newcomers can exploit the system of differences to their advantage while drawing various forms of capital.

The Currency of Cultural Capital

I had a chance to interview Alarcón in his offices at Columbia University, where he teaches journalism. It is important to know that Guerrero and Alarcón are not radio people, at least not in the conventional sense. That is to say, they had no formal training in public radio before starting a podcast that would, within just a few years of its founding, be carried by NPR.

Before co-founding the podcast, Guerrero worked as a promoter for cultural and social projects, creating a bridge between organizations in her native Colombia and public and private institutions in Latin America and the United States, designing and managing festivals and art exhibits, teaching workshops, and planning fundraising events (Women's Leadership Accelerator 2017).

Alarcón is, first and foremost, a novelist. When we spoke, Alarcón described how he finds it interesting that these days he is known primarily as a radio person. "As a writer and a novelist, it's always surprising to me that people are like, 'oh wait, you write novels too?'" he told me. "Because my core identity, like the pillar of my identity since I was nine was like, I'm going to be a writer, and so I think it's funny that people are like, he's a radio person that also writes novels."

Alarcón was born in Peru and raised in Birmingham, Alabama, in a solidly middle-class family. His parents, both physicians, sent him to a progressive private high school (Rohter 2013). He attended Columbia University, where he received a degree in anthropology. Eventually, he made his way to the University of Iowa's Writer's Workshop, which led to his career as a novelist (Rohter 2013).

Alarcón speaks both Spanish and English, but, in many ways, he exemplifies Weinrich's ideal bilingual. His ability to write and speak capably in Spanish and English has enabled him to navigate multiple spaces throughout his career. He can host a Spanish-language radio program, but also can serve as the de facto spokesperson for the program in English-language media. But it becomes clear that bilingualism and biculturalism are central to Alarcón's identity. "I think I'm an American writer writing about Latin America," Alarcón stated during an interview in the *New York Times*, "and I'm a Latin American writer who happens to write in English."

When we met, Alarcón described how, even though he wasn't formally trained in radio, the medium has been important to him. It was a way to make connections with his family in Peru: "My family would often make these kinds of radio programs on cassette for our family back in Peru, that we would send by mail. And to anyone under a certain age, that sounds completely insane. This was before the technology we have now that allows us to be in constant communication with our families around the world. So that's the back-back story."

Alarcón has told this story often in interviews. He is, after all, a storyteller. And as a good storyteller, this ongoing connection to radio establishes a narrative thread that links his early life to his literary work to his current enterprise. When discussing his early life, he described his family as a radio family:

> My dad was a soccer announcer in Peru, and I have
> cousins, and uncles and family members who worked
> in community radio stations and small radio stations all
> over Peru. And when I was a kid, we listened to NPR a
> fair amount. Obviously for the news and information and
> all that. And also, because my father, my parents, were
> very interested in us getting the NPR accent, in English.
> To them, it sounded like the educated accent, you know.
> And certainly, I've sort of inhaled that.

Alarcón's description of his household as an NPR listening space reminded me of something Ray Suarez, host of NPR's *Talk of the Nation*, told me when we spoke about the kinds of Latinxs who can enter the NPR realm. According to Suarez: "To be an NPR kind of person, it helps to have grown up in a household where NPR was playing. NPR speaks with a cosmopolitan, educated, upper-class voice. It's someone that's interested in new cuisines, who travels, someone will listen to a piece on world music. It's a cultural location and a class location more than an ethnic location."

Largely due to his success as a novelist, Alarcón certainly possesses cultural capital, which has served him and the program well. Alarcón has become a darling of the educated elite. The *New York Times* called him "a writer that thrives in two cultures" (Rohter 2013). He has written for a variety of publications in Peru and the United States. *Granta* named Alarcón one of the best young novelists under the age of thirty-five, and the *New Yorker* listed him on the "20 under 40." In 2005 he wrote *War by Candlelight* (2006), a collection of short stories. A year later his first novel, *Lost City Radio* (2007), was published.

The book is, fittingly, about a public radio program—one that is broadcast out of the capital city of an unnamed country, but likely a stand-in Peru. The title *Lost City Radio* refers to a fictional call-in show where listeners hope to reunite with their loved ones who have been displaced or have disappeared during the country's dirty war. The protagonist of the book is Norma, host of the program, whose job it is to read the names of the disappeared men, women, and children from the city and from the outlying villages and settlements. The radio program becomes a way to give voice to the voiceless, to build a community of listeners, and to connect listeners from different parts of the country, ranging from the city to the settlements.

The success of *Lost City Radio* led to an opportunity to participate in the production of an audio docuseries for the BBC, which touched on one of the themes in Alarcón's book—the migration from the villages to the city. Alarcón told me about this experience:

> That year I was asked to do a radio documentary by the BBC. They sent a producer from London to Lima. I flew to Lima and we did this hour-long documentary on Indian migration to the capital, which is something that I'm very interested in. So, we went to a region called Ancash which is where my grandfather's from. And we visited this town that was emptied out, and we visited this community in Lima, and we did this radio documentary.

The experience gave Alarcón meaningful exposure to audio production, but he soon realized that there were profound creative deficiencies once one eliminated the actual voices of those storytellers. He told me:

> I did interviews in English and in Spanish. The producer,
> who came from London, spoke no Spanish. So, I served
> as simultaneous translator. The documentary, obviously,
> had to be in English, so what ended up happening is
> that there was this, in the final edit they cut most of the
> voices that I found most compelling. Those were voices
> in Spanish. And used voices in English. I'm proud of the
> documentary. I think it's pretty good. I also think it's, sty-
> listically, not what I would have done. And I also felt like
> it was really frustrating on an aesthetic level, on a jour-
> nalistic level, to not hear those voices. So, I was left with
> the question. What would it be like to have this space in
> audio for the kinds of stories we want to tell?

According to Alarcón, the question lingered for a few years, but the idea for *Radio Ambulante* began in earnest over a conversation at a coffee shop in the bay area, where he and Guerrero lived at the time. The idea was to produce a long-form radio show in the spirit of *This American Life* and *Radiolab*, but they recognized that they had neither the skills nor the reputation in radio that were necessary. Furthermore, because they were on the outside of public media, they did not have access to economic capital, which would typically include funding from the CPB or the foundations that subsidize public media.

But what Guerrero and Alarcón lacked in professional experience, they made up for in cultural and social capital. The team there has been able to develop strong collaborations that have compensated for professional and financial limitations. But the first step was to test the idea itself:

> I remember the first discussions of it. We said, oh, we
> should write an e-mail to some friends. And I had worked
> at this magazine called *Etiqueta Negra*, which is kind of
> this legendary Peruvian magazine that was founded in

Lima by a friend of mine named Julio Villanueva Chang. I served as the American editor of that magazine. And that magazine became kind of the extended family and the staff and the folks around it became the first wave of collaborators for *Ambulante*.

According to Alarcón, his connections to writers in Latin America were extremely beneficial in the early days. "Things like that were incredibly helpful and it also helped to be part of the literary community, you know. And to have a lot of connections there." Guerrero also brought in her professional connections. According to Alarcón, "Carolina brought a ton of Colombian folks that she knew and that we had met here in the states. First generation, Latino mostly, who were interested in this kind of storytelling."

Once they validated the idea, they started to build a team of journalists and producers. The group brought in Martina Castro as a senior producer, and she designed most of *Radio Ambulante*'s sound (*Sounding Out* 2014). Castro had started her career at NPR and brought significant experience to the project. I spoke with Castro about those early days, and she described what it was like to launch an upstart podcast at the time. According to Castro: "Podcasting at that time remained this rebel, and we were like, 'we're going to do something new.' There were no gatekeepers. And there wasn't Spanish language anything in public media. We thought, 'you need to break through and do it on your own.' And we really felt like we were doing that. Applying lessons that I learned at NPR obviously."

According to Castro, the team wanted to develop quality storytelling in Spanish. To accomplish this, however, the team needed to find a balance between respecting the particularities of a given region while telling a story that would have universal appeal: "A lot of people questioned whether a Latin American podcast would be successful. That Mexicans would be interested in a story based in Argentina, and vice-versa. . . . Borders don't exist for us. We're just all human beings. We all saw the potential where we could celebrate our differences. We never want to white-wash anything or paint with a broad brush."

According to Alarcón, the fourth member of that original team was Anna Correal, a journalist from Colombia, who is also a native Spanish speaker: "We reached out to this woman named Anna Correal who works at the *Times* now. She was a reporter at *Diario*, which is a Spanish language newspaper. And she had done this piece for *This American Life*. And she's just a kick-ass reporter. Just a beautiful writer. Anyway, she was one of the first people to come on board, and she actually came up with the name."

The name *Radio Ambulante* refers to the mobile street vendors who are ubiquitous throughout Latin America, and this idea is also built into the program's logo, which depicts a silhouetted man pushing a food cart in the shape of a vintage radio. According to Guerrero, they focused on these vendors because, in her words, "these are super-brave, resilient people, who walk, travelling around the city, moving all over the public space; they are persevering. We felt that this image reflected us particularly well. And we loved the idea of our logo, of someone carrying a radio" (Guerrero 2020).

The next step was raising capital for production. According to Fernández-Sande (2014), Alarcón invested $25,000 of his own money to launch the project and the creation of the two pilot programs. However, when they failed to win any foundation grants, he decided that crowdfunding was the logical next step. At that point, the team had the idea for the program and a sampler with fewer than forty-five minutes of audio (*Sounding Out* 2014).

There is a fundraising video called "Welcome to *Radio Ambulate*" that was produced as part of the team's Kickstarter campaign (*Radio Ambulante* 2012). Shot in both English and Spanish, the video has low production quality and poor sound—a world apart from the slick, highly produced videos of established news organizations. In it, Alarcón, Castro, Correal, and Guerrero explain the idea behind the program and its basic working model. The narrative behind *Radio Ambulante* is all there: the novel, the BBC docuseries that followed, the need for stories that feature Latin American stories in their own voices. All of it. But then comes the pitch to listeners. "We're asking for your help to make this project a reality. We're currently producing our first three pilot episodes," Alarcón tell prospective funders. "We need to purchase our equipment, our production costs, and support our hardworking producers in over a dozen cities."

The fundraising campaign was a success. During the January through March 2012 fundraising period, six hundred donors pledged a total of $46,032, which was above the target of $40,000. After commissions were deducted, they had $41,000—enough to seed the project for a few months (Fernández-Sande 2014). *Radio Ambulante's* expenses included salaries of its core staff and external collaborators, commissions paid to commissioning journalists, costs related to the design and maintenance of the *Radio Ambulante* website, and the purchase of technical equipment.

Once they got the seed money, they then went on to produce their pilot episode. Alarcón discussed the first year as challenging and humbling. I asked what made that first year so difficult:

> Learning what the hell we were doing. I'd written long-form journalism, I'd written for magazines like *Harpers* and *The New York Times* magazine, and the *New Yorker*, and *Granta*. I mean, I took the tools of creative non-fiction and reported narrative non-fiction, whatever you want to call it into my toolbox. And it was something that I always enjoyed doing. But always in print. And so I had to learn how to do it in radio. And if you listen to the first season of *Radio Ambulante*, well God bless you, because I think it's not that good. It's ambitious, but I think there's a lot of growing up in public.

Alarcón has written about the first experiences producing a podcast in an essay titled "Storytime at the Azteca Boxing Club" (Alarcón 2017a). In it, he describes travelling to Bell, California to interview Kina Malpartida, a legendary Peruvian boxer. Malpartida kept Alarcón waiting for over two hours, and, when she finally arrived, was only half engaged. This encounter was not enough to make for an interesting podcast. But, while he was waiting, he began a conversation with a Peruvian immigrant named Mayer Olórtegui, who had gone to see Malpartida train. The two began to talk, and it became apparent that Olórtegui had some interesting stories of his own.

The next day, he interviewed Olórtegui, which became the story for "Los Polizones" (2012), one of *Radio Ambulante's* first podcasts. The podcast tells the remarkable story of Mayer Olórtegui, a Peruvian who was from Callao, just outside Lima. In 1959, Olórtegui and a friend emigrated to the United States, sailing as stowaways in the hold of a ship, when Olórtegui was only nineteen years old. But, as Alarcón tells it, capturing Olórtegui's story was an unexpected surprise. On one hand, the essay is about being open as a journalist to finding stories, but, on the other hand, it's also a candid picture of what it's like to learn audio-storytelling:

> But how do you go from an idea, or a set of principles, to an actual show? In practical terms, we were fumbling in the dark—which is the best way to figure it out, of course. You must fumble boldly and with purpose. That's what I was doing in Los Angeles at the Azteca Boxing Club, wasn't it? You do an interview. What next? Select the best tape. Bang out a script. Read it aloud with your cuts. Your heart sinks. So, you make something new, and hope that this time it's better. It may take a dozen attempts or more, but eventually, it does get better. It's magical. This is the process, and there are no corners to be cut. Every mistake, every shitty script, is a small down payment on future excellence. (Alarcón 2017a)

According to Alarcón, it took them about six months to produce their first episode, and then another four months to produce their next two. As he points out, the first stories were simple, often using single voices and straightforward language. The first two seasons also cover territory that would have been familiar to Alarcón, with a focus on interviewing novelists and journalists, including Junot Díaz and Francisco Goldman (2013), Yuri Herrera (2012), and Guadalup Nettel. Over time, however, the show would diversify the kinds of stories it told and would also benefit from partnerships with English-language podcasts.

Courting National Public Radio

Like many fields, public radio is an industry of connections. One of the contacts who proved to be most instrumental was Mandalit del Barco, who was an arts and entertainment reporter at NPR, but who was brought in as a senior advisor for *Radio Ambulante*. According to Alarcón, it was del Barco who brought the idea to NPR. "Anyway, she brought our content to NPR. I remember we had a call, maybe our second season," Alarcón recalled. "And NPR passed. They were like, this isn't really for us."

I asked Alarcón why he believes that NPR decided to initially forego a partnership with *Radio Ambulante*. Alarcón was candid in his assessment that the quality of the program needed to be improved, including his own performance: "It was mostly because the host sucked, basically, is what they said. And I had to take it on the chin. And in some ways they were right. So, some of the concerns that I had mentioned earlier about my tracking, specifically, early on. They weren't wrong, to put it mildly."

Here, Alarcón is referring to the difficult transition writers often make between print and audio. Alarcón reported that he had experience reading his work all over the world. But serving as the host of a podcast was different. This is not necessarily surprising. The participants with whom I spoke discussed the same difficulties. As one voice coach who has consulted with NPR stated, "many [contributors] come from print journalism and then join NPR. They need to get comfortable. What works for print is not the same as what works for the ear."

But these are matters of execution. I asked Alarcón if he believed that NPR bought into the concept itself—the idea of a Spanish-language, long-form narrative podcast.

> Not in any real way, no. I think there was a certain amount of deference and respect for a veteran [journalist] like Mandalit. Like, "we have to take the meeting because Mandalit has some pull around here." So, I think there was that. I don't think that there was any commitment, like, we're definitely interested in that. So, they passed. We briefly signed a deal with PRI. But we were basically, for all intents and purposes, independent. We had a very, very small budget. Funded by small foundations and we would teach workshops here and there to help pay for things. But it was very, very small.

Alarcón is describing a short-lived collaboration with PRI. With a grant from PRI's New Voices Fund, the team produced English-language stories for PRI's *The World* (Lapin 2013). After developing as a program and building new relationships, however, they took the project back to NPR. By then, two things had changed. First, they had evolved and had collaborated with other well-known podcast producers, which showcased their ability to produce a program that was NPR quality. Second, they found an unexpected advocate for the program at NPR—in this case, Lynette Clementson, who served as senior director of strategy and content initiatives. According to Alarcón:

> There was an executive at NPR, she was relatively new, named Lynette Clementson. Lynette had helped co-found *The Root*. And had been a national reporter and had done a ton of great stuff. And she came in and being an African American woman, part of her purview was, let's look at diversity in a real way at NPR. What I recall is that she built an ad hoc group of reporters and staff of color and would ask them, what can we do? What can we do to change the dynamic and make a real impact?

Alarcón credits Clementson with creating a space where a cohort of people of color within NPR could brainstorm pathways for new entrants. But it was also important that *Radio Ambulante* demonstrate its ability to produce a quality program that would appeal to NPR listeners. Here, partnerships became an important factor:

> By that time, we had published a story with Radiolab, called "Los Frikis," and it was about Cuba. It was a co-production and we had done a version in Spanish, and they had done a version in English. That was a huge inflection point for us. Because what happened was that Adrian Florido, actually, went to Lynette and was like, you should listen to this. And it was in English, so she could get it. And she was like, the people that do that, that's the kind of stuff that we want. So, she reached out to us and we started a long process of negotiation. And eventually signed a three-year deal to be distributed by NPR.

Cuando La Habana era Friki (Alarcón 2017b) is a podcast that focuses on a Cuban punk band whose members injected themselves with HIV in order to find reprieve from government oppression in a sanitarium during Cuba's Special Period in a Time of Peace. The project was produced in partnership with *Radiolab*, a podcast produced out of NPR member station WNYC and distributed by NPR. But, because of the partnership, the producers of *Radio Ambulante* were able to demonstrate that they had the same journalistic and production standards as NPR.

Eventually, NPR signed on as exclusive distributor, and, according to Alarcón, the relationship between NPR and *Radio Ambulante* has been mutually beneficial. First, *Radio Ambulante* benefits from NPR's marketing and public relations apparatus. The formal announcement for the partnership was preceded by an NPR marketing push for the show at the Third Coast International Audio Festival conference. Referred to as "the Sundance of radio," the conference is meant to showcase emerging audio storytellers.

NPR also brought some support from the Ford and McArthur foundations (Guerrero, 2018). Between 2017 and 2019, the program was awarded $600,000 in grants for journalism and media. The show has also benefited from synergies within NPR's product portfolio. For example, *Radio Ambulante* recorded "¿Qué le pasó a José de Jesús?" (Alarcón 2016), which was essentially a shorter retelling of *The Strange Death of José de Jesús* (2016), in Spanish in collaboration with *Latino USA*. The following year, Alarcón sat down for a recorded interview with Felix Contreras (2017), host of NPR's *Alt.Latino* podcast, and, for half an hour, Alarcón was able to describe the program, its mission, and the process by which it generated stories. In doing so, the podcast was able to gain exposure to *Alt.Latino*'s own audience.

But, ultimately, Alarcón believes that *Radio Ambulante* has benefited from the symbolic value that NPR brings to the program. According to Alarcón, their association with NPR serves as a form of legitimization as an upstart enterprise:

> It's a seal of quality because *Radio Ambulante*, literally for years, was Carolina and me and a few colleagues in our basement. A lot of foundations were very interested and intrigued. Everyone thought it was cool and they could see that our audience was growing. Still, they were like, "Hey, how do I know it's any good?" Then NPR is on there and they're like, "OK, it must be pretty good." That's

a huge thing. When we were looking for partners and there were other organizations that were interested in partnering with us, it came down not so much to money but to shared journalistic values.

According to Alarcón, this kind of institutional legitimacy is especially important for a Spanish-language, Latinx-/Latin American-focused program, which would otherwise escape the notice of industry leaders.

NPR brings brand glow, I guess, is one way to think about it. This stamp of approval. These upstart brands who do things in Spanish and nobody knows if it's any good . . . this stamp of approval from NPR allows you to go to foundations. And that has certainly paid off. I mean foundations have certain eyes for us than they did before. And the other thing they do. If we can get on the shows, like *Morning Edition, All Things Considered, Weekend All Things Considered*, then we have exposure to an incredible audience.

By partnering with a large, well-funded, and reputable organization, *Radio Ambulante* stands to benefit from NPR. I asked Alarcón how NPR, in turn, benefits from the relationship, to which he responded:

A partnership like the one we have with NPR. What do we bring? We bring this audience. We bring this amazing content. We bring access and knowledge about Latin America that is useful for the newsroom. I'm frankly connected to journalists across Latin America that you can call for whatever, you know? Like the rolodex that Carolina and I, and the project has, would be the envy of anyone trying to cover Latin America. And we have that.

Certainly, the potential upside for NPR is high. With relatively little investment, NPR can cultivate a large audience beyond its national boundaries while making inroads with U.S. Latinx listeners. In addition to expanding its reach, NPR can alleviate concerns that it is not investing in diversity programming. Within NPR discourses, *Radio Ambulante* is put forward as evidence that NPR is serving the Latinx community.

Journalists Without Borders

According to Fernández-Sande (2014) *Radio Ambulante* represents a new paradigm for radio journalism and production that is aligned with today's digital economy. The team at *Radio Ambulante* has been able to create a thriving brand by taking full advantage of digital convergence and the multiple platforms now available for dissemination of news. Over the years, the team at *Radio Ambulante* has hosted a number of lectures, interviews, and workshops about radio production. Consequently, the process by which *Radio Ambulante* produces their stories has been well documented.

It becomes readily apparent that *Radio Ambulante* does not look like a traditional newsroom with a centralized office. *Radio Ambulante*'s offices are located in New York City, but their core team includes people working from New York, Bogota, Xalapa, Quito, San José, London, and Hamburg. This kind of team resembles what Leslie (1995) describes as a "transnational business community," which is based less on fixed spatial concepts and more on transnational flows of information and capital.

Digital technology guides all facets of the process, including production, creative development, and distribution. Developments in communications technology mean that a small upstart team has been able to produce high-quality production with relatively inexpensive equipment. According to Correal (2011), the team began by investing in production quality, digital recorders, and microphones, which they used to experiment. They then loaned the equipment out to journalists in other countries. They have also used digital platforms as a way to distribute that content worldwide. Before they had a partnership with NPR as exclusive distributor, *Radio Ambulante*'s producers were loading the finished piece onto SoundCloud, an online audio distribution platform. Finally, they use communications technology to access talent globally.

In a lecture at the Drescher Institute for the Humanities, Guerrero (2018) detailed the process by which the team produces their stories. Staff meetings are held on the video teleconferencing platform Zoom, as well as Slack, a platform for organizing communications for group discussions by channels that allows for private messages to share information, files, and more, all in one place. The team also uses Google Drive and Dropbox for sharing and editing files, and Trint, an automated transcription service (Guerrero 2018).

According to Guerrero, the process of creating a podcast can take between two and six months. In some cases, investigative pieces can take much longer. The first step is story selection, where the team receives ideas for stories from all over the world. At that point, they will conduct some preliminary research to verify some of the details of the pitch, then evaluate the story and make sure that it is a good fit for *Radio Ambulante*. According to Guerrero, one of the characteristics they are looking for is an element of surprise, meaning it is something that moves beyond a conventional story.

Second, they are looking for stories that are character driven. The general topic may be about a larger social issue, but the podcast generally focuses on a particular individual's story. For example, a podcast titled *El Fotógrafo* (Alarcón 2017c) focuses on Victor Basterra, an Argentine photographer who was taken prisoner during the country's dirty war. The story painfully traces his abduction, imprisonment, and torture. He was then persuaded by his captors to serve as a photographer in part of a scheme to falsify documents. These photographs would later be used as evidence during prosecution efforts. While the focus is on Basterra, the podcast is really about the larger issue of the military dictatorship and the scores of Argentines who were disappeared during Argentina's dictatorship.

If the team decides that they want to pursue a given story, they will assign an editor and producer. However, instead of tapping into an existing community of public radio professionals, the producers at *Radio Ambulante* have cultivated a professional community of Latin American print journalists, which means they come to the project with different professional sensibilities. Because these journalists have not been groomed in the public radio system, they are not beholden to the conventions of public radio and are, therefore, open to more innovative techniques. But it also means that time must be invested in training these journalists to write and produce for long-form, narrative, audio storytelling.

Alarcón reported that, particularly in the early days of the program, none of the journalists with whom they collaborated had experience with audio, which meant that they had to train them in the basics of radio production—recording, logging their tape, and writing a script. In an interview with *Sounding Out* (2014), Alarcón argued that technical aspects of gathering sound are less important than conducting a good interview, which is the basic building block of a good radio story. He urged journalists to phrase their questions so that they could "get vivid, almost filmic answers, full of details that set the scene" (*Sounding Out* 2014).

Once they get the interview material from the journalists, *Radio Ambulante* will then transcribe the audio and create an outline with the help of an editor. Before a script is final, it is shared with other editors on the team around the globe. Postproduction includes several rounds of editing, as well as music production and the sound mixing of the various voices within the story—including the hosts and actualities. According to the group, the goal is to showcase the many characters involved in any given story, while giving the listener enough contextual information to immerse themselves in the story.

Language and Linguistic Markets

Given the format, the producers at *Radio Ambulante* pay particular attention to voice. As host, Alarcón plays an important role in moving the story along without becoming too much of an interfering presence. Furthermore, the podcasts often use multiple characters, so the producers find ways to clearly demarcate speakers. For the producers of *Radio Ambulante*, however, it is important that listeners hear the voices of the people actually involved in the story. Therefore, journalists are encouraged to write scripts that highlight this intimate connection with the listener: "We ask our reporters to write colloquially, to imagine they're telling the story to a friend at a bar. It's important to have immediacy in the language, an expressive tone that can seem almost improvised, even when it isn't. The emotional impact of radio is that it feels as though a secret is being shared. The script and the production should always be in service of this intimacy" (*Sounding Out* 2014).

I found this to be an interesting variation on NPR's trope of the "listener as friend." As I have previously argued, NPR's imagined friend is a highly educated English monolingual, and distinctly American. But *Radio Ambulante*'s ideal listener is much more broadly defined. Given the reach of their program, that imagined friend could be from a small town in Colombia or a borough in New York. Given the geographic span of the program, Spanish serves as an important unifying framework. During an interview with Nieman Labs (Nieman Reports 2020), Guerrero described the need to attend to Spanish sensitively: "In our case, the kind of journalism that we do is very personal, so we use Spanish. It would lose too much in a voice-over. When I grew up listening, for example, to the BBC on the radio, I would hear a voice-over, not the voice of the protagonist."

In a public talk, Guerrero (2018) described how Spanish may be a unifying framework, but the team makes a concerted effort to highlight the linguistic diversity across the Americas, stating that "it's not like we wanted to unify all the accents and the cultures. Actually, it's just the opposite. We wanted to make sure *Radio Ambulante* makes clear what is different and what is similar."

There is a promotional piece that Alarcón and Guerrero have shown during their public presentations, called "Las Voces of Radio Ambulante" (2014). It is essentially a two-minute sound montage of some of the actualities in their various stories. In it we hear the voices of speakers from Puerto Berrío, Colombia, Havana, Cuba, Brooklyn, New York, Ville Alemana, Chile, Cordoba, Argentina, Montevideo, Uruguay, and Tegucigalpa, Honduras.

It is an effective piece of communication that complicates the notion of Spanish as a monolithic linguistic identity. Instead, we hear distinct accents, cadences, and intonations, ways of speaking that show the diverse range that exists within the Spanish language. However, there are limits to this range, and it is important to consider who is left out. None of the podcasts are produced in Mapuche, Triqui, Quechoa, or any of the other indigenous languages that are widely spoken throughout the Americas. The focus on Spanish, in other words, obscures the lack of linguistic variety that persists within Latin American countries despite globalization (Garcia-Canclini 2001). I asked Alarcón how language figures into how they imagine their ideal listener:

> Our audience is . . . the easiest way to talk about it is a pie that's divided into thirds. And the numbers vary from story to story and season to season, but it's been steady, when we had an audience of 700 and when we had an audience of 120,000 per episode. Roughly, it's still the same. So, a third would be Latinos and Latin Americans in the U.S. So that means everything from the first-generation kid, the one and half generation who prefers to speak in English but understand Spanish fine.

Alarcón then described the other listeners:

> Then there are people like Carolina, who moved to the states to work or to study in their twenties. That's one broad group, okay? So, what you call heritage speakers and Latin American migrants. Then you would have Latin Americans in Latin America. That's another third. And our top countries are Mexico, Colombia, Peru, Argentina. And then the third would be Spanish-language learners. Non-Hispanic, Non-Latino Spanish-language learners. So, folks like college students who are learning Spanish from study abroad. Folks who are our most committed probably audience.

In interviews, Alarcón has described this audience as a "three-legged stool" (Lapin 2016) that includes a broad coalition of U.S.-born Latinxs, new immigrants, and Latin Americans. But, upon further reflection, these listeners will have vastly different experiences, linguistic practices, and regional affiliations. I asked Alarcón if there was any connective tissue that unites these disparate groups: "Yeah, I would say that the connective tissue is an appreciation for stories. An interest in Latin Americans and Latinos. An appreciation of the wide spectrum of accents and vernaculars that are spoken across Latin America. I don't think I have an ideal listener honestly. I think I'm always impressed and astonished by the range of folks that listen to us. And glad for it."

Alarcón's claim that there is no ideal listener seems counterintuitive. After all, *Radio Ambulante* delivers a consistent product with a consistent editorial point of view. As a media product, it will appeal to some taste publics and not others. Furthermore, the team would certainly have needed to identify a target audience when pitching prospective foundations and media partners. It becomes clear that this listening community closely resembles what Garcia-Canclini (2001) refers to as a "transnational consumer community," one defined less by territorial limits or the political history of the nation-state and more as an interpretive community of media consumers. As such, these listeners reflect what Appadurai (1996) refers to as a cosmopolitan sensibility, or a global ethos characterized by an openness to cultural experimentation, hybrid identity, and international cultural transfer and exchange. During our conversation, for example, Alarcón recalled a moment encountering one of these listeners in New York:

> We had a crazy thing happen this winter, when Carolina and I were outside this bar in Brooklyn. We were waiting for an Uber or something. And we were talking in Spanish about this event we had just been at. And some guy comes up to us. This, like, prototypical hipster. And he was like, "hey I'm sorry, but I couldn't help but hear your voice and I was wondering are you Daniel Alarcón from *Radio Ambulante*?"

For Alarcón, it was a moment of realization that the program was reaching just more than Latinxs and Latin Americans: "And I was like, that's crazy. Because when we started out, we didn't think we would be a white, Brooklyn hipster product, but I think it's great. I think it's great because people . . . if we can tell stories and make people curious about Latin American and Latino communities and bring up important issues and tell important stories too. I think that's only a plus."

That *Radio Ambulante* would find a loyal listener in a white, Brooklyn hipster, however, seems entirely likely, given that *Radio Ambulante* delivers a taste public that fits nicely with NPR's target audience. The notion of the "curious listener" has pervaded NPR's discourses throughout its history. It is a term that suggests inclusivity but that appeals to listeners who have access to circuits of global knowledge, culture, and information. From this vantage point, the United States is simply another nodal point in a much larger network of Spanish-speaking (and even Spanish-learning) listeners, tied together through their media consumption practices.

The goal of appealing to the right kinds of listeners was built into the original vision of the program. In an article written for *Transom* (2011), Correal described the ways in which Guerrero, who has a background in arts management, researched the opportunities that existed within the "Hispanic market," which is itself an industry construction that reorganizes publics according to their economic capacity. According to Correal, Guerrero gravitated toward "the Fusionista," someone living in the United States who is fluent in Spanish and English, but who lacks good content in Spanish.

Correal is likely referring to an audience segment that was invoked by Hispanic advertising agency Alma DDB, in their report titled "A Brave New World of Consumadores" (2011). In it, the agency describes a market segment they call Fusionistas, or "new Latinos," who are English speaking but have a dual cultural affinity. According to the agency, this audience segment prefers to speak Spanish at least half the time. Most importantly, they are also prolific users of Spanish-language radio. Their usage of Spanish radio is over 60 percent, which is nearly twice their consumption of Latinx-oriented television, magazines, and newspapers.

But, as the report makes clear, it is their importance as consumers that is paramount, stating "this whole new demographic of young, reasonably affluent Hispanics entered the world of marketing as incredibly important and valuable consumers" (Alma DDB 2011). By courting this consumer, the producers at *Radio Ambulante* have demonstrated an innate ability for entrepreneurship. Over time, the team has figured out ways to monetize the product by building partnerships with both nonprofit and commercial organizations. In short, the team has been able to align their social mission with their economic interests.

This was particularly evident during *Radio Ambulante*'s launch of *Lupa*, a mobile app that they began promoting in 2019. *Lupa* is a collaboration between *Radio Ambulante* and Jiveworld, an e-learning company based out of Mill Valley, California. The app segments the *Radio Ambulante* podcasts into short lessons, where users can adjust the difficulty level by slowing down the audio and by swapping back and forth between the Spanish transcript and its English translation. Users can also highlight specific words and look up their meaning. Those words are added to a vocabulary list that they can study later. The app is designed so that, over time, the user relies less on the English-language transcript.

Investing in an app that cultivates a Spanish-speaking audience is, in some ways, a political act. As I have previously discussed, over time NPR has become a monolingual network that has pushed out other languages. But *Lupa* is one of the few examples that I found that actively challenges the standardizing practices at NPR by encouraging the network's listeners to learn Spanish. At the same time, the app is a revenue-generating mechanism. At $99 a year, the project also brings in some revenue to *Radio Ambulante*. *Lupa* is one of several initiatives designed to the grow the business.

Radio Ambulante also has gained success by entering into new partnerships with commercial media. In 2020, they launched *El Hilo* (*The Thread*), a weekly Spanish podcast of news and analysis, which was developed in collaboration with Acast, a global podcast company. The newsfeed was designed to fill the void left with the closure of *New York Times en Español*. *El Hilo* features interviews with journalists and regional sources to present the week's news stories from a Latin American perspective. They say the podcast "is designed to deliver stories in a style similar to *Radio Ambulante*, but grounded in a newsier approach" (Quah 2020).

Radio Ambulante's effective use of language to build an audience reflects a movement that del Valle (2006) calls the concept of La Hispanofonía, a movement started in the second half of the nineteenth century. It was the spread of the idea of a pan-Hispanic brotherhood and was based on the belief that, in spite of the political independence of the former colonies, a Spanish culture embodied in the Spanish language remained as an inalienable link among all Spanish-speaking nations. The new paradigm has triggered the breakdown of the traditional relationship between the nation and its language and rearranged the distribution of value to different types of linguistic competence. Thus, language is reconstructed as a commodity.

From this perspective, producing stories in Spanish is an economic endeavor. The producers have been upfront in saying that they have been able to exploit the system of differences within public radio to their advantage. In a saturated marketplace, Spanish can serve as point of difference. According to Guerrero, "We can see now that the opportunity is there for us, because the market in the U.S. is a bit saturated with English language podcasts. A podcast is created like every day, like one hundred of them. Instead, there are not that many podcasts in Spanish." In this way, they operate a lot like dominant Spanish-language media organizations, which see language as a point of difference that must be protected.

Thus, the *Radio Ambulante* listener can be seen as a member of a cultural-linguistic market (Straubhaar 1991), one that is geographically dispersed but culturally and linguistically linked. At the same time, English-language podcasts do not see them as competitors, which means they can partner with English-language podcasts such as *Radiolab* that are also seeking to expand their listener base. This brings some value to NPR, which is exploring ways to extend its brand into new markets. During the launch of their partnership with *Radio Ambulante*, NPR's senior vice-president of programming, Anya Grundmann, touted the unique reach of the podcast: "*Radio Ambulante* brings us sound-rich stories from all over North and South America. NPR is so proud to support the most ambitious, deeply reported, nuanced Spanish language podcast. Their work adds range and depth to our journalism and highlights voices and perspectives we don't normally hear."

Behind Grundmann's elation, there is a bit of doublespeak at play. In an essay titled "NPR in Spanish: Approaching Content for a Bilingual Audience" (Pretsky 2017), Grundmann touts NPR's support of *Radio Ambulante* while simultaneously affirming the network's position that NPR is, in her words, "basically an English-language American broadcaster." So, even though the network will support the program in podcast form, Grundman is clear that its Spanish-language content will not be a part of the NPR terrestrial experience.

When they began their partnership with NPR, *Radio Ambulante* had hoped that their stories could be featured on NPR's flagship programs, *All Things Considered* or *Morning Edition*. However, this has not happened as much as they had expected, which Alarcón attributed to unfortunate timing. *Radio Ambulante* launched during the 2016 election cycle, and Alarcón believes that since then the news cycle has pushed out the more evergreen content that *Radio Ambulante* produces.

We also discussed complications when it comes to negotiating the practicalities of the relationship with the leadership at NPR, who are primarily English monolinguals. I asked Alarcón how senior managers at NPR can evaluate the work itself if they cannot understand it. The process of evaluation can be challenging:

> The process of evaluation, as you've noted, is always difficult. We do great work. How do the folks in charge know that? How do we prove it to them? Collaborations with English language projects help. Often there are individual staff members, who are tasked with listening and evaluating. We've never had anyone in a leadership position who speaks Spanish or even understands. We're always in a position of advocating for ourselves and proving to others that what we do is quality.

The team recognizes that, to promote their brand, they need to engage English-monolingual gatekeepers, and they have made some efforts to ensure some presence in English language media. For most of the podcasts, the producers will develop English-language transcripts, which make the content comprehensible to English-dominant speakers. Furthermore, as a member of the literary community, Alarcón has access to dominant news organizations, which enables the team to generate additional awareness of their stories. In select cases, variations of their stories have been written for English-language publications.

Storytelling in the Age of Trump

Radio Ambulante began their partnership with NPR in 2016, the same year Trump took office. Like other journalists, producers have had to confront the administration's xenophobic policies and anti-Latinx rhetoric. I asked Alarcón how this reality has impacted his show and its reception at NPR:

> A couple of things. I think there's something inherently political about producing a show in Spanish in this political climate. And the rhetoric around Latinos that we're

all intimately familiar with. That's something that can't be denied. Built into the DNA of *Radio Ambulante* is the premise that the United States is a Latin American country. And that on its face . . . and that's not to say that it's not an Asian American country too and it's not an African American country. It's just that one of the things we believe and is one of our principles is the idea that there is another America living in this country, that prefers to speak Spanish. That doesn't mean they're not American. That the "public" in public radio, sometimes speaks Spanish. I don't think that's a controversial statement, but I can see how some people would think it's a controversial statement.

The idea that the United States is a Latin American country is a provocative statement, one that is meant to complicate the narrative that U.S. Latinxs are residents of some other place. But Alarcón makes an essential point that Latinxs are members of the public that NPR is tasked with serving. Alarcón believes that it is essential to reinsert Latinxs into the national dialogue. He went on:

That's one way of addressing the current situation. Is by simply existing. Having said that, I think that our audience expects, quite rightly, that we're going to cover immigration. And we do. And we cover it extensively and we cover it sensitively. And we cover it from the point of view of the people living it, primarily. And so, we rarely have expert voices. I don't want a professor telling me what it's like to be raided by ICE. I want the workers that were there to tell me. And the families that were most affected. The kids. And I think that's a really important difference. Not necessarily different from shows like *This American Life* which covers immigration that way, but from the news shows on NPR.

This perspective was evident in an episode called "Una Cadena Humana," (Alarcón 2019), which centered on a community in Hermitage, a neighborhood of Nashville, Tennessee. In July of 2019, a father left for work with his twelve-year-old son when ICE agents blocked his pathway and ordered him out of his van so that they could detain him. The agents had no official warrant, and, without a warrant, they could not forcibly enter the family's car or their home. When the man refused to leave his car, the officers then attempted to coerce him out by threating to detain his family.

Then, unexpectedly, the neighbors and activists in the community began to mobilize. First, in small ways. They made signs in support of the family, such as one that read "We're with you. We love you." Because it was a hot summer day in Tennessee where temperatures can be oppressive, the father and son were comforted by their neighbors who gave them something to eat and drink. But then it escalated. The neighbors gathered together to build a human chain, surrounding the boy and his father and then moving along with them so that they could safely enter their home. After several hours, the ICE agents left.

Essential to the story, however, is the power of voice. We hear from a variety of participants, in English and in Spanish. We hear, untranslated, the testimonies of Latinxs who are directly involved in the process, not simply passive agents. We hear from journalist Verónica Salcedo of *Nashville Noticias*, a digital media outlet geared toward the Spanish-speaking community. She received the call for help from the family. We also hear from Cathy Carrillo, a member of MIX, an organization of young people that seeks to combat racism and improve immigration laws. During the podcast, Carrillo tells her own story of her father's deportation one afternoon while driving to one of her sporting events. Immigration officials detained her father, who was eventually deported to Peru. She was abandoned on the side of the road.

This kind of storytelling provides important human insights into U.S. immigration policy, which, the team feels, is an especially important need during the Trump administration. Yet, like other Latinx journalists, this practice subjects them to accusations of bias. Alarcón rejected this framing: "I think when you have people who are advocating for the separation of families. Who are quite specifically advocating causing pain to others as a policy goal in order to affect one outcome or another. I think there's only one way to label that and that's cruelty."

In 2018, the team at *Radio Ambulante* felt compelled to directly address these issues. That year, the team decided to produce an extra episode called "Tolerancia Cero" (Alarcón 2018), which centered on

Trump's child separation policy. Unlike *Radio Ambulante*'s other stories, which focus on a particular character's story, the episode takes a macro-level view with details on Trump's policy. But it also differs from U.S. reporting, which is much more U.S.-centric. The podcast takes a transnational view by providing insight from Honduras, where many of the immigrants targeted by Trump hail from.

Compared to their other podcasts, this episode is a much more pointed critique of Trump's immigration policy. Furthermore, there is a clear tone of indignance about human rights abuses that differs significantly from the cool, dispassionate voice that is characteristic of NPR's news programs. The producers at *Radio Ambulante* believe that such an approach, if done well, can provide clarity. "I think we try to report on that cruelty in a way that creates empathy and understanding for the people who are living it," Alarcón told me. "And I think that's a perfectly valid journalistic stance."

ALT.LATINO

The sky plane caught fire over Los Gatos Canyon,
A fireball of lightning that shook all our hills,
Who are all these friends, all scattered like dry leaves?
The radio says, "they're just deportees."

— WOODY GUTHRIE, 1948

On January 28, 1948, the U.S. Immigration Service chartered a World War II surplus DC-3 from Airline Transport Carriers of Burbank to transport twenty-eight migrant workers from Oakland to the deportation center of El Centro, California. Some of the migrants were part of the government-sponsored Bracero program, while others had come to the United States without documents (Marcum 2013). Somewhere over the ranches on the edge of the Diablo Range, twenty miles west of Coalinga, the plane began trailing black smoke. Rancher W. L. Childers reported that the plane was at about five thousand feet when he noticed the smoke from the left engine of the two-engine plane. Then the wing fell off, and the fuselage plummeted, nose first. As it fell, Childers said several of the passengers either jumped or fell to their deaths. There were no survivors (Hernandez 2017).

The next day, the crash received national news coverage, in which the crew was identified as Frank Atkinson, a pilot from Long Beach; his wife, Bobbie Atkinson; and Marion Ewing, the plane's co-pilot (*Los Angeles Times* 1948). The Immigration Services guard was identified as Frank E. Chaffin of Berkeley. When it came to identifying the Mexican passengers, however, the reports failed to name those who perished, instead describing them collectively by their immigration status. Both the *Los Angeles Times* and the *New York Times* quoted Irving Wixon, director of the Federal Immigration Service, who described the deceased: "The group included Mexican nationals who entered the United States illegally, and others who stayed beyond duration of work contracts in California . . . all were Agricultural workers" (the *New York Times* 1948).

Moved by the incident, folk singer Woody Guthrie wrote a poem that called attention to the dehumanizing practices of the news media, which rendered the migrants invisible. In it, Guthrie reimagines these migrants as individuals who are leaving their friends and lovers to travel to the United States to grow crops, only to be humiliated. "You won't have a name when you ride the big airplane," Guthrie wrote. "All they will call you will be deportee." Titled "Deportee (Plane Wreck at Los Gatos)," the poem was later set to music by schoolteacher Martin Hoffman and became a favorite of folk singers and social activists, having been covered by artists including Pete Seeger, Judy Collins, and Joan Baez.

Fifty years later, on July 11, 2018, a remake of the song was referenced in a forty-five-minute special edition of NPR's music-oriented podcast, *Alt.Latino*, titled "Protesting Trump's Immigration Policy Through Song" (Contreras 2018). The podcast was produced specifically to address the Trump administration's policy to separate families at the border by showcasing the efforts of various musicians, working across genres, who were drawing attention to the issue.

Early in the show, Felix Contreras, the show's producer and host, highlights the ways in which artists are using music as form of protest. According to Contreras, there is a connection between Guthrie's song and more recent efforts by musicians from across the Americas to call attention to social injustices faced by U.S. Latinxs:

> Singing about social conditions has a long tradition, with artists such as Víctor Jara and Ruben Blades in Latin America, and Bob Dylan and Woody Guthrie in this country. In fact, in 1948 Woody Guthrie actually wrote about the Mexican immigrant experience in a song about a plane crash in Central California that killed twenty-eight migrant farm workers being deported back to Mexico. . . . The song has been covered by many artists since then, but here's a favorite of mine, from a new album by Los Texmaniacs featuring Lyle Lovett on the vocals.

In some ways, Contreras is doing what Guthrie himself was attempting to do, which is to provide a human face to those who are often erased by the national media. In the process of drawing attention to Trump's policy, the podcast features songs that span the emotional spectrum. There's a collaboration between Austin-based artist David Garza, and Paulina Reza, a member of the Mexican indie band The Chamanas. The duo recorded a cover of the ballad "Besame Mucho," which strikes a melancholy tone, recorded with each artist performing from a different side of the U.S.-Mexico border. Then there is guitarist Marc Ribot's experimental cover of "Rata de dos Patas" ("Rat with Two Feet"), originally sung by Mexican artist Paquita la del Barrio. The song is defiant, with lyrics focused directly at the president: "filthy rat, creeping animal, scum of life, deformed monstrosity, sub-human specter of hell. How much damage you have done?"

Alt.Latino's pointed critique of Trump's immigration policy is a significant departure from NPR's general news coverage, which has historically failed to capture the full humanity of U.S. Latinxs. For example, in her analysis of NPR's coverage of Latinx immigration, Noel (2017) found that, despite their efforts to produce content for a highly educated and civically minded listener, NPR reifies larger discourses that frame U.S. Latinxs as social and economic threats. The day before the *Alt.Latino* podcast launched, the administration's child-separation policy barely made news on NPR. A headline for NPR's *Morning Edition* online appeared to frame the issue as an administrative mishap, reading "Government misses Migrant Family Unification Deadline" (Burnett 2018).

Alt.Latino demonstrates the capacity for producers working for NPR to negotiate and even subvert the network's institutional norms. However, the ability for *Alt.Latino* to act as a form of resistance is predicated on the presumption of access and agency. In her research on community radio, Castañeda (2016) found that Latinxs have historically faced economic and institutional barriers to accessing airwaves, undermining their ability to serve local communities and engage with national issues that affect Latinxs. However, access is not necessarily a guarantor of social change. Cultural producers from marginalized groups often work within an institutional logic that frames the work they conduct on behalf of media institutions, limiting their ability to enact social change (Smith-Shomade 2008).

In this chapter, I explore the kinds of oppositional work in which the producers of *Alt.Latino* are engaged, as well as the limitations of these efforts within the framework of NPR. My focus in this book has been primarily on news, but music provides a unique space in which to explore racial politics. I begin by describing the ways in which various forms of Latinx protest and musical expression have been circulated within the framework of the commercial music industry. I follow by describing public radio's ambivalent relationship with commercial music, which has packaged Latinx music as an educational rather than a commercial product. Finally, I examine the economic conditions and the logics of practice that shape *Alt.Latino*'s creative output, including its conception of an ideal Latinx listener. Based on interviews with the show's creators, I argue that *Alt.Latino*'s designation as a "music" show enables it to engage in oppositional work not possible in NPR's general news programming.

Latinx Music, Protest, and the Commercial Music Industry

In 1930, two Mexican brothers, who performed under the name Los Hermanos Bañuelos, recorded a song titled "El Deportado" for Vocalion records. The duet had received some recognition four years earlier with their cover of "Lavaplatos," a corrido about an immigrant who seeks success in Hollywood but finds only work as a dishwasher and returns to Mexico with little to show for his experience (Schmidt Camacho 2008). "El Deportado" covers similar territory. Sung in two parts, the song tells the story of a man who leaves Mexico for the United States, escaping the violence of the Mexican Revolution. In Part I, the narrator documents his journey by train, passing the towns of La Piedad, Aguascalientes, and La Chona. Throughout the song, the narrator expresses the sorrow of leaving his mother.

In Part II of the song, the narrator describes life in the United States, singing about the indignities he faces while in the country. He arrives at the border, where immigration officials make him feel worthless, describing him as dirty and poor. The white people he meets in the United States are cheats, taking advantage of his situation. As the song concludes with his deportation, he laments the treatment of Mexican migrants in the United States:

Adiós paisanos queridos	Goodbye my beloved countrymen
ya nos van a deportar	we are being deported

| Pero no somos bandidos | But we are not criminals |
| venimos a camellar | we came to work hard |

In 2013, the Los Angeles-based band La Santa Cecilia released a song titled "El Hielo (ICE)," referring to the Immigration and Customs Enforcement agency. In the song, the band similarly addresses the risks that undocumented workers in the United States face every day, where they live under constant threat of deportation. Like "El Deportado," the song also drives home the ways in which forced separation leads to the devastation of Latinx families:

El Hielo anda suelto por esas calles	ICE is loose all over the streets
Nunca se sabe cuando nos va tocar	We never know when we'll be touched
Lloran los niños lloran a la salida	They cry, the children cry at the doorway
Lloran al ver que no llegará mama	They cry when they see that their mother will not return

Despite the eighty-three years that separate the two songs, there are striking similarities. On one hand, both songs demonstrate how little has changed over the past century when it comes to immigration policy. Both songs were produced by Los Angeles-based artists who are addressing the indignities of being undocumented in the United States.

On the other hand, there have been profound changes in the music industry, which have transformed the ways in which music is produced and consumed. It is important to recognize that "El Deportado" was sung in corrido form, a narrative ballad, which began as an oral tradition in the last half of the nineteenth century (McDowell, Herrera-Sobek, and Cortina 1994). The corrido involves the telling of a story, and generally ends with a farewell (Paredes 1958). In its earliest forms, the nature of the song would change, depending on the singer and the context (Paredes 1958), but this tradition shifted at the turn of the century, first with the mass production of sheet music, thereby fixing language, and later with recorded music, in which a song could be attached to a single voice or artist.

By the time Los Hermaños Banuelos recorded "El Deportado," large record labels were emerging, primarily with the strategy of producing music to sell phonographs. While white, middle-class consumers preferred nationally popular and classical music (Frith 2006), music companies found that they could also make money by producing and distributing vernacular and regional music. All the major record labels developed separate labels that acquired and distributed African American and Latinx music, or "race records" (Sánchez 1995). These included RCA Victor's Bluebird Records and Columbia's Okeh Records. Vocalion, the record label under which "El Deportado" was recorded, was a subsidiary of Brunswick Records.

In the American Southwest, these labels recognized Mexican buying power and collaborated with Latinx middlemen, who identified talent from across the Southwest. A discography of records produced during that time shows they recorded songs in Los Angeles and San Antonio (Spotswood 1991). Sessions were recorded with a portable recording machine in a hotel or makeshift studio, and musicians were generally paid between $15 and $20 for a session. These records would then be sold for a profit. In an effort to sell more records, many of these songs were recorded in two parts.

These records were often sold through Latinx-owned music or furniture stores and marketed to Latinx consumers with specialized appeals. In his analysis of early Latinx advertising in Los Angeles, Chávez (2014) describes a full-page ad that ran in a 1926 issue of *La Opinion*. Placed on behalf of Repertorio Musical Mexicano, a small business specializing in records, sheet music, and phonographs, the advertisement features the store's logo, which consists of an image of a vaquero strumming a guitar and sitting atop of a large phonograph. The copy reads "La única casa de música Mexicana para los Mexicanos. Establecida en Estados Unidos para Los Mexicanos." (The only house of Mexican music for Mexicans. Established in the United States for Mexicans.)

By the time La Santa Cecilia released "El Hielo (ICE)" in 2013, there was tremendous consolidation within the industry. Currently, Universal Music Group (UMG), Sony Music Entertainment, and Warner Music Group, known as The Big Three, account for almost two-thirds of all music sold worldwide (Resnikoff 2016). All have Latin American divisions and vast marketing channels throughout Spain and the Americas. "El Hielo (ICE)" was recorded under the UMG label, which cultivates a global Latin audience market from its bases in Los Angeles and Miami.

Within the industry, La Santa Cecilia is marketed as a Latin Alternative band, a genre that itself has an ongoing history. In his research on the music industry, Corona (2017) describes the evolving ways in which Latinx musicians have engaged rock music. Originally, Latinx artists, including Ritchie Valens and Los Teen Tops, de-emphasized ethnicity in favor of a homogenous rock identity, but Latin Rock, denoting a new style blending rock, R&B, blues, and jazz with Caribbean origin and Latin rhythms, began to acquire currency in the late 1960s. At the time, Latin Rock was defined by its otherness, meant to appear as essentially dissimilar to mainstream (white) rock in its introduction of "foreign" sounds and ethnic markers.

The industry's shift from *Latin Rock* to *Rock en Español* was significant because it placed less emphasis on U.S. Latinx artists, thereby capitalizing on rock's popularity in Spain and Latin America. In doing so, the genre embraced Spanish as the primary language. At the same time, both categories reify a parallel musical universe, framing it as a music market for Latinxs by Latinxs. When *Alt.Latino* first aired on NPR in 2010, its producers were building on the music industry's growing interest in Latin Alternative, music that combines new music styles (hip hop, electronica, rock, etc.) with traditional artforms.

The commercialization of Latinx music is said to leave marginalized communities at a distinct disadvantage by reducing people's historically specific cultural experiences to objects of consumption (Feld 2000). However, some scholars also see possibilities within this framework. The same channels of investment and commerce that exploit specific communities also carry their music to a wider world audience. In doing so, they allow marginalized groups to develop a shared sense of community, which can then create new political possibilities. For example, Kun (2005) argues that music can act as a sort of transnational borderland in which U.S. and British rock and pop styles fuse with traditional and modern artforms from music of the Americas. These spaces provide musicians and listeners with the opportunity to challenge the commonsense sound of what it means to be American.

NPR's Ambivalent Relationship with Latinx Music

As Kun (2005) argues, none of this is happening outside the marketplace. The commercial, global capitalist channels of the recording industry have allowed Latinx artists to remap national identity. However, public radio reflects a very different relationship with Latinx artists. As a noncommercial enterprise, public radio has traditionally promoted ethnic music as an educational rather than a commercial product, which has resulted in a high culture/folk binary that has defined NPR's music strategy over the years.

At the onset of radio, popular music was seen as emblematic of what was wrong with commercial radio. The rising popularity of ethnic music was seen as evidence of radio's debasement of national culture. For example, Adorno's (1945) critique of the broadcasting radio was partially in response to jazz, an ethnic artform that he believed to be standardized and uncritical relative to classical music, which he viewed as a superior form of music.

Seen as an alternative to commercial radio stations, proponents of early educational radio during the 1930s believed that radio could elevate the masses by introducing them to classical music, fine art, history, and politics (McCauley 2005). It appears as if listeners gravitated toward educational radio stations for this reason. When the *Hidden Medium* report was published, the authors found that music was one of the primary reasons that audiences listened to and supported educational radio. According to the report, "although the interest in other programming keeps growing, music remains the preferred material: classical, semi-classical and lighter music, in that order" (Herman W. Land Associates, Inc. 1967, I-18).

When NPR launched, the network inherited educational radio's mission of noblesse oblige. NPR's original mission statement reflected this goal of promoting high culture. In it, Siemering (1970) argued that symphonies should no longer be limited to the elite, but rather made available to the masses: "Art is no longer a pleasant pastime for a social elite, but at the core of contemporary life. Understanding is both more essential and more difficult. . . . National Public Radio might use any of the following to make the arts understandable and engaging."

To accomplish this, Siemering believed that NPR might stimulate local orchestras through national broadcasts, rather than focusing primarily on larger, established institutions like the New York Philharmonic. Instead, there could be a concert series with a different local orchestra each week. But Siemering also wrote that NPR could use music as a way to enact its diversity mandate through music, writing that "a sense of the cultural diversity could be achieved by programs featuring the music of the different ethnic groups across the country" (1970).

Almost immediately, however, classical music became the default on NPR at the national level. When NPR first aired, its first broadcasts included recordings of the Los Angeles Philharmonic (Lochte 2003). This was partially a function of NPR's economic reality at the time. For an upstart network with relatively little funding, classical music is cheaper to acquire, thereby enabling the network to deliver music programming without incurring the substantial licensing costs. However, during the network's early history, classical—and, later, jazz—became a core component of NPR's brand identity. In an effort to affirm its musical identity, NPR has marketed a series of CDs and books, including *The NPR Curious Listener's Guide to Classical Music* (Smith and Thomas 2002), *The NPR Classical Music Companion* (Hoffman 2005), and *The Curious Listener's Guide to Jazz* (Schoenberg 2002).

By contrast, NPR's support of regional and ethnic music was left primarily to local member stations. On how NPR might actually promote the music of various ethnic groups, Siemering's vision for NPR was vague at best. "Programs 'by and for' specific cultural, ethnic minorities category could be developed," Siemering (1970) wrote. "For example, there could be a linkup of stations in urban areas with sizeable non-white audiences, or student groups studying ecology, or groups with distinct lifestyles and interests not now served by electronic media."

Locally, stations were experimenting with programming formats in which different music programs were aimed at different audiences, what Bailey (2004) describes as a "Crazy Quilt" strategy. However, the AUDIENCE 88 report promoted two changes that would essentially shape NPR's music strategy both nationally and locally. First, there was a clear shift away from arts and culture to news programming. Second, local stations were encouraged to avoid a "diversity strategy" and instead concentrate their efforts on their core audience of educated, affluent listeners. As the network began to rely more heavily

Alt.Latino

on audience data so that it could tailor its programming to its most lucrative listeners, NPR's managers began to conclude that focusing primarily on classical and jazz music was no longer a winning strategy, because it appealed to a small but not necessarily loyal audience (Ferguson 2004).

For NPR, the presentation of ethnic music has served as a form of ethnomusicology in which an imagined listener was educated on various cultures through music. At the national level, NPR has assumed the music industry's overall practice of characterizing Latinx musical styles as ethnically distinct from normative, Western, European music. Because Latinx music is framed in terms of its otherness, NPR takes on the additional labor of explaining the music—and the cultures they represent—to an imagined audience of primarily white, English-monolingual listeners.

Consider how NPR categorizes various kinds of Latinx music in their *Curious Listener's* book series, which instructs its audience on how to appreciate NPR's music. Various music genres, such as Conjuntos, Corridos, and Mariachi, are categorized by NPR as American folk (Lornell 2004), which is defined as "music with strong regional affiliations or a distinctive racial/ethnic identity as well as direct links with the past" (Lornell 2004, 1). Conversely, contemporary bands including Los Aterciopelados and La Mano Negra are identified both as "Rock en Español" and "World Music" (Nickson 2004) because they perform in Spanish and are not of U.S. origin.

The slippage between Latinx music, folk music, and world music is consistent with the recording industry's overall practice of distinguishing Latinx music as distinctly ethnic and, therefore, essentially different from the (white) mainstream (Corona 2017). By categorizing Latinx music as the music of some other time or place, NPR reifies what Corona (2017) describes as the semantic boundaries used to police various musical genres, a process that involves a number of players, including corporations, producers, marketers, critics, and consumers.

Over time, NPR has transformed itself with the shift from analog to digital. As Harvey (2014) observes, in the new broadcasting environment, listeners enjoy more access, personalization, and choice compared to the traditional mass-media paradigm. This has changed the role that broadcasters play in determining music tastes. Cwynar

(2017) further argues that, in the new radio landscape, human curators provide a degree of authenticity and a point of identification for users.

NPR capitalized on this paradigm shift by launching "NPR Music," an internet-based service that intends to appeal to users who wish to experience music through NPR's curatorial sensibility. This effort has been a success. NPR has evolved into what Paul Farhi (2010) of the *Washington Post* called "a taste-making force in the fragmented music business." In doing so, NPR exploited the overlap of "public" and "indie" culture, which frames itself as an "autonomous, authentic alternative" to mainstream commercial culture, despite NPR's commercial nature and the fact that the network serves the interests of a class of sophisticated consumers who hold significant amounts of cultural capital (Newman 2009).

This is the context in which *Alt.Latino* first aired on NPR. Unlike *Latino USA* and *Radio Ambulante*, *Alt.Latino* is one of the few Latinx-oriented programs that is produced by NPR. In recent years, the network has turned to podcasting as a way to diversify its audience (Powell 2015), and NPR's support of these programs signals some promise for Latinx cultural producers, working within the framework of NPR, to fulfill the public radio's promise of delivering relevant programming for Latinxs. From this perspective, it is important to consider the industrial conditions that enable counter-hegemonic discourses to occur on NPR as well as the limitations of these practices within the framework of public radio.

The Oppositional Work of *Alt.Latino*

Originally broadcast in July, 2010, *Alt.Latino* showcases Latinx artists while addressing current political, social, and cultural issues. Each podcast centers on a particular theme. For example, podcasts have been built around such diverse topics as the FIFA World Cup, coffee, and telenovelas. *Alt.Latino* also produces its annual *Sonic Altar* series, which runs in celebration of Dia De Los Muertos. To help guide the listener through the experience, *Alt.Latino* often integrates the perspectives of guest DJs, who include a range of Latinx public figures, including musicians, journalists, and scholars.

Originally, *Alt.Latino* was meant to fill a clear organizational need for NPR, which was to help diversify its audience by drawing in Latinx listeners. When I spoke with Contreras, he told me that funding for the program was initially made possible through the 2012 National Endowment for the Arts sponsored by the Arts on Radio and Television Fund and a Sterling Clark Foundation grant. The specific directive of the grant was to reach out to listeners underserved in the media marketplace. According to Contreras: "We had been messing around with the pilot . . . and NPR put out the request for proposals because they had money for new podcasts and they wanted to reach out to underserved audiences."

It appears as if the network was soliciting projects that would deliver on what CPB chief content officer, Joseph Tovares (2016), would later describe as the CPB's strategic Three D's: digital, diversity, and dialogue. That is, any new projects that would expand NPR's operations into digital broadcasting, diversify its audiences, and engage listeners in new ways.

According to Contreras, the idea for *Alt.Latino* emerged out of a series of informal conversations he had with Jasmine Garsd, who, at the time, was working as a production intern. As Contreras reported:

> Jasmine Garsd was a producer for this show, *Tell Me More*.
> And we had struck up a friendship. She's from Argentina,
> but we had struck up a friendship based on the fact that we
> were some of the few Latinos in the building. And then we
> had this common appreciation for music. Latin Alterna-
> tive music. Basically, what became *Alt.Latino*. At the time,
> it was considered Rock en Español. Hip Hop en Español.
> All these different genres.

When I spoke with Garsd, she also recalled being one of the few Latinxs at NPR, which helped to create a bond with Contreras, who, by then, had a long-established career in radio and television. "He worked right by the snack machine," she recalled about meeting Contreras, "but we'd talk about music. It was this really interesting connector. He was this Mexican American guy from California. And then you had this Jewish, Argentine immigrant in her twenties."

Garsd would later go on to work as a senior reporter for American Public Media's *Marketplace*, but she recalled feeling a bit disaffected during those early days at NPR. She believes this disaffection impacted how she thought about *Alt.Latino*, which allowed her to work through her experiences of being a Latinx journalist in a primarily white, professional space. According to Garsd: "At the time, I was very cranky and felt alienated. And at the time, I couldn't put it into words. What was happening to *Tell Me More*. It felt very alienating in the way things are covered. At the time, I didn't have the vocabulary to understand it."

Garsd is referring to the eventual cancellation of *Tell Me More*, an interview show hosted by NPR's Michel Martin. The show was credited with addressing issues of diversity, but its cancellation after seven years sparked outrage from some of its fans, who claimed that NPR was walking away from issues important to listeners of color (Schumacher-Matos 2014). Garsd felt that NPR's approach to Latinx-oriented issues was symptomatic of a larger problem in the media, where representations of Latinxs can either be ignored or framed in reductive ways. Garsd believed that *Alt.Latino* could be a space where Latinxs could talk about these complexities. According to Garsd: "A lot of it came from personal experience. What I've found over time. I just feel alienated by the reductive representation of ethnicity in this country. Whether you're a gay, Mexican American or an Afro-Latino, a Jewish South American. You feel very weird about these categories. When I came to this country. I felt I was alone in my alienation."

For Garsd, *Alt.Latino* was envisioned as a shared space where Latinxs from a wide range of backgrounds could share their collective experiences. In this way, the producers of *Alt.Latino* conduct oppositional work by resisting what Garcia-Canclini (2001) calls the "frivolous homogenization" of diverse cultures by the marketplace. A stated goal of the producers is to disrupt limited perceptions of Latinidad by showcasing the diversity of the Latinx population. During an interview with *USA Today*, Garsd asserted the following viewpoint:

> *Alt.Latino* matters because Latinos matter, and because we aren't getting our due. Latin culture is so varied, so rich. But it you look at the media landscape, the little attention we do get tends to be this really reductive, simplified, often stereotyped version of Latin culture. This is changing, and we're excited to be part of it. We love focusing both on what unites us as Hispanics, and what

makes us a really varied group that is hard to categorize. (as quoted in Alvarado 2013)

As a music-oriented show. *Alt.Latino* is in a unique position to serve as a platform in which to showcase the racial, ethnic, and linguistic diversity of the Latinx community. The inclusion of musicians, particularly Afro-Latinx musicians from the Caribbean, helps the producers to enact this goal. For example, when *Alt.Latino* showcases the music of Dominican hop hop artist Ariana Puello, they are also providing a platform for her perspective:

No soy morena, soy negra	I'm not dark, I'm Black. Learn it!
¡apréndetelo!	
No te equivoques conmigo	Don't make that mistake with
¡recuerdalo!	me. Remember it!
Oye racista ignorante, asúmelo!	Listen, ignorant racist. Face it!

In "Asi es La Negra," Puello is not only asserting Black pride, but she is also overtly challenging listeners' preconceptions of what it means to be Latinx. Visually and sonically, the show's guests diversify what has traditionally been a white public space. Images of the musicians and the hosts, in the form of headshots and photographs taken in the recording studio or on location, are posted on the *Alt.Latino* microsite, making it one of the more visually diverse spaces within NPR's website.

Garsd and Contreras also believe that they embody a more complex notion of Latinidad by defying visual stereotypes. Garsd, the immigrant, is light skinned, while Contreras is darker and American-born. According to Contreras: "The premise of the show, by the way, was that you had this young person from Argentina, who was an immigrant, bilingual, Spanish-dominant. And then an older, Chicano born, second-generation, born here in the United States. English is my dominant language. That's the whole point. This is America. This is the Latin community right now. These two different things."

Sonically, the show also pushes against NPR's standard approach to language. According to Garsd, she and Contreras also made a conscious effort to move away from the NPR standard language, which she felt was suppressive of ethnic voices, stating, "we both really wanted to move away from that clean, sanitized, NPR sound that gets spoofed on SNL.* We both grew up listening to Latin radio and Black radio. And we wanted that bombastic, clanky sound."

It becomes clear that the show brings a noticeable linguistic diversity to NPR. As the network has increasingly affirmed its commitment to English monolingualism, *Alt.Latino* is one of the few spaces on NPR where listeners might encounter consistent use of Spanish or code-switching. According to Contreras, they have experimented with different ways of expressing themselves, including Spanish, English, or some combination of both. Most of *Alt.Latino*'s podcasts are produced in English, but significant portions are broadcast in Spanish. This is largely a function of the artists featured on the show, many of whom are not native English speakers As Contreras stated, "we've had the luxury of experimenting with language, but eventually falling back on 'it's an English language show and we can bring in Spanish as needed, when people are expressing themselves.'"

Contreras' description of *Alt.Latino* as an "English-language show" provides some insight into how language is treated at the network. For example, Garsd discussed her experience interviewing Spanish-dominant musicians. "Take Mala Rodríguez," Garsd told me. "She doesn't speak much English, so we interviewed her in Spanish." Garsd is describing a 2011 interview with the Spanish rapper (Contreras and Garsd 2011). During the interview, Rodríguez responds in Spanish, but her testimony is then translated online for an audience who does not speak Spanish. Listeners are then provided a link that reads "escucha el programa en español," where they can listen to the untranslated interview. While some of the podcasts are available in all Spanish, Contreras acknowledges that the member stations that carry *Alt.Latino* will not run them.

To a lesser degree, the show has also showcased indigenous languages in an attempt to stretch the Spanish–English binary that has come to exemplify Latinidad in the United States. This is best illustrated in the podcast episode titled "Hear 6 Latin American Artists Who Rock Indigenous Languages" (2013). In her introduction to the episode, Garsd challenges listeners to resist normative conceptions

---

*In 1988, *Saturday Night Live* spoofed the NPR voice with a series of sketches that centered on a fictional show called *NPR's Delicious Dish*.

of indigenous language, stating that is "something that's labeled 'old world music' or ethnic music, which is stored away to gather dust, but something that was alive and dynamic, that included electronica, heavy metal, reggaetón" (Garsd 2013). To this end, *Alt.Latino* showcases indigenous languages in new ways. The songs featured in the podcast include a club remix of Bolivian artist Luzmila Carpio, who sings in Quechua; the Guatemalan band, Lago Negro, who integrate the Mayan language Tz'utujil into their music; and Palin Aukatuaiñ, a collective that raps in Mapuche, the indigenous language from Chile and Argentina. I asked Garsd where the idea for this program originated. She reported that it was a reflection of what was happening at the time on the U.S. border: "At the time, there was a whole crisis that began to boil up. People were arriving at the border who spoke Mayan languages. Amerindian languages. In Argentina, there's a lot of oppression and xenophobia, but here it's suffocating and claustrophobic. There's this concept of what it means to be Latino that gets pushed on us. It's suffocating."

By drawing connections among these disparate groups within the Americas, the show reflects a cosmopolitan sensibility (Appadurai 1996) that is characterized by a global ethos and transnational cultural exchange. *Alt.Latino*'s primary focus on music enables the producers to more easily facilitate these exchanges. As Lipsitz (1994) argues, music creates opportunities to build solidarity with other marginalized groups through exchange, and even appropriation. Similarly, Shank (2014) argues that music can create a shared sense of community among marginalized listeners that can result in a collective identity that can, in turn, create possibilities for political change.

I found that the show's producers have made a conscious effort to directly address political issues. In his critique of NPR, Loviglio (2013) argues that the network reflects strong neoliberal sensibilities. However, I found that *Alt.Latino* reflects more progressive politics. A number of the podcasts take a critical view of global capitalism and offer a sympathetic view of those left out. For example, past podcasts have focused on the Dreamers (A Song for the Dreamers, 2018) and farmers (Songs of Hope and Struggle, 2016). Others have focused on gay pride (Love and Pride on the Dance Floor, 2016) and sexual empowerment (La Mostras: Fierce Women of Latin Music, 2016).

The musicians themselves become the conduits for addressing these issues. But they do so in a way that transcends regional specificity. For example, when *Alt.Latino* features songs by Chilean singer Ana Tijoux, she may be singing about student demonstrations in Chile, but in interviews Tijoux has described how her music also speaks to the Arab Spring and Occupy movements (National Public Radio 2012). In a similar way, the show has showcased the song "Latinoamérica" by Puerto Rican band Calle 13, which describes a shared history and collective strength while critiquing Western force and influence within Latin America (Garsd 2012).

The show's political orientation, however, is most evident on the topic of immigration, which has been a consistent focus. Over the years, the producers of *Alt.Latino* have taken a critical view of U.S. policy during both the Obama and Trump administrations. According to Contreras:

> Early on we wanted to address immigration. And we thought about, how do we do this? Understanding that we are a product of NPR. And NPR is primarily a news organization. We can't take a stand. We can't say that what we think is happening at the border is wrong. We have to have some sense of objectivity. The idea was with immigration, we decided to do a series of programs that dealt with immigration, but in different ways. The first one we did was "Songs about the Immigration Experience."

The podcast episode to which Contreras is referring is titled "5 Songs About Immigration, in Song Form" (Contreras and Garsd 2013), which was meant to provide some context for the national debate on immigration reform. The podcast featured everyday Latinxs who discussed their varied, often frustrating, experiences with the immigration system. In the process of telling their stories, these individuals referred to music as a way to articulate their emotional state of being. At one point, a woman identified as Roxanna Sánchez described the feeling of desperately wanting to belong in a country that she believes does not want her in return. To Sánchez, the song "Frijolero" (Beaner), performed by the rap band Molotov, gives voice to her frustration: "Pensando en tu familia mientras que pasas dejando todo lo que conoces atrás . . . las seguirás diciendo 'good for nothing wetback.'" (Thinking of your family while leaving everything you know behind . . . they'll still say "good for nothing, wetback.")

That Sánchez can find solace in Molotov, a band from Mexico City who are singing about racial politics in the United States, exemplifies the kinds of global solidarity promoted by the podcast. In similar ways, Mexican regional music has been used to respond to the current U.S. immigration policy. At one point during the podcast, a woman identified as Claire Hidalgo describes what it is like to be a U.S. citizen married to someone who is not. Given that the mobility of undocumented immigrants is highly restricted, something as small as boarding an airplane becomes an unthinkable act. To Claire, the song "Si Nos Dejan" (If They Let Us), by the Ranchera artist José Alfredo Jiménez, gives voice to her frustration:

Si nos dejan	If they let us
nos vamos a vivir a un mundo nuevo.	We're going to live in a new world.
Yo creo	I believe
podemos ser un nuevo amanecer de un nuevo día.	We can be a new dawn Of a new day.
Yo pienso	I think
que tú y yo podemos ser felices todavía.	That you and I can still be happy.

Both Garsd and Contreras indicated that their personal and professional experiences inform how they think about *Alt.Latino*. Before coming to NPR, Contreras worked in Spanish- and English-language media as a journalist at NBC and Univision. But as a college student at Cal State Fresno, he also worked as a volunteer at Radio Bilingüe, which, at the time, was an upstart radio station. According to Contreras:

> In 1980, one of my college professors had started a radio station, he said he was going to start a radio station. It was going to be a listener supported bilingual station. It was the early days of Radio Bilingüe in Fresno and Hugo Morales was my college professor. I was taking this media

class. I was a broadcast major. I offered to volunteer. So, I was one of the first five volunteers at Radio Bilingüe when it went on the air in 1980.

Garsd described a different pathway. According to Garsd, her family left Argentina in the midst of economic turmoil, which she believes shaped both her perspective and her desire to be a journalist: "The experience of my country collapsing. We came to the U.S., we lived in a hotel. I worked a ton of jobs, along with a lot of undocumented workers. And those experiences made me think, I want to do something more tangible."

Contreras argues that he can use music as a platform to draw attention to social issues of the day. "We can't make a statement," Contreras stated, "but we can reflect what's going on in the world, either through interviews, or making some kind of music theme about it. . . . I'm going to show musicians who are reacting. [These artists are saying] 'we're using our art.'" As Contreras suggests, *Alt.Latino*'s designation as a music show allows the producers to do oppositional work that is not likely to be done in NPR's news-oriented programming.

Here, Contreras is pointing to the distinctions among various journalistic genres, which predetermine how the show will be read by its intended audience. As McKee (2003) argues, genres work by providing conventions that allow efficient communication between producers and audiences. Knowing genre and its rules helps producers to make reasonable guesses at how a text is likely to be read by audiences. As a music-oriented program, the ways in which *Alt.Latino* addresses social issues will be read differently than if they were to be covered on NPR's "hard news" programs such as *All Things Considered* and *Morning Edition*. Understanding these rules affords the producers some flexibility to take a more pointed approach toward these topics; however, there are limitations to this practice. Music-oriented programs, however political, are not placed on the same level as "serious news" within the journalistic field, and there is little expectation that systemic issues will be addressed in detail. Consider Garsd's more recent work as a senior reporter for *Marketplace*, a radio program that focuses on the U.S. economy. In this role, Garsd has been able to address, in more explicit ways, the economic inequalities that profoundly impact Latinx labor, health, and education.

In his discussion of NPR's targeting practices, Loviglio (2013) argues that, since its inception, NPR has engaged in an ongoing struggle to reconcile two competing audiences. On one hand, the network was created with the expectation that it develop programming for disenfranchised communities, which had traditionally been left out of civic discourses—what Loviglio describes as the "under-served, under-educated citizens at risk" (Loviglio 2013, 142). On the other hand, implicit racial ideologies (Garbes 2017) and economic necessity have motivated NPR to target its programs toward an almost exclusively upscale audience of white, educated baby boomers (McChesney 2008).

These tensions become manifest in *Alt.Latino* through its conception of the audience, as well as its design. As Garsd told me, "the biggest negotiation we had with producers was around the question of audience. Felix and I wanted the audience to be Latinx, but I think management wanted an audience who we could explain Latino culture to." Both Garsd and Contreras indicated that they wanted to produce a show that would appeal specifically to Latinx listeners, which they believed was necessary given NPR's conservative approach in the past. According to Contreras:

> I don't have a problem saying it. It speaks to public radio's inability, slash, failure to reach underserved audiences. They, as a whole, have a hard time reaching, in particular Black and Latino audiences. I've been here since 2001, in the building at NPR and that's always been the goal. More diversity. Like they say, "behind the mic and in front of the mic." But they're just not taking any chances.

Contreras went on to describe how NPR's current programming strategy can seem out of touch with working-class Latinxs, who may currently be listening to Radio Bilingüe. According to Contreras, there are Latinx listeners who may be predisposed to public radio, but, when they go to NPR, they do not find programming that is relevant to their lived experiences:

If you have a Radio Bilingüe audience but you turn on NPR and hear an interview, week after week, an interview with white authors. Or somebody doing some obscure off-Broadway play. Which are two very real examples. And I think, what does this have to do with anyone else beyond this east coast white mentality? It's almost like, okay NPR, you've got to blame yourself because look at what you're programming. How are you bringing those people in? They hear people talking, but they're talking about things that have nothing to do with their lives. *On Point, Here and Now*, any of those shows. They have themselves to blame.

Garsd expressed a similar sentiment, arguing that NPR's current programming strategy has impeded its ability to appeal to Latinx listeners: "It's like if you had a house party, and you've only invited white people for years. And then you want your Latino friends to come . . . you're gonna have to put out some flyers. If you want new people to come on board, you're gonna have to try new things. Introduce some new sounds. You want something new."

Both Garsd and Contreras believed that *Alt.Latino* could be a space where NPR's current listeners and Latinx listeners could coexist, but the first step was identifying which kind of Latinx listener they wanted to reach. According to Contreras, the proposal for *Alt. Latino* was informed by a Pew study that showed that the highest users of mobile media were Latinxs aged eighteen to thirty-four who are English-dominant bilingual. Contreras is likely referring to a study conducted by the Pew Hispanic Center titled "Latinos and Digital Technology" (Livingston 2010), which found that 81 percent of native-born Latinxs are online, compared to only 54 percent of foreign-born Latinxs. At the time of the study, English-dominant and bilingual Latinxs had relatively high rates of internet use. Of all English-dominant Latinxs, 81 percent were online, as were 74 percent of bilingual Latinxs. In contrast, only 47 percent of Spanish-dominant Latinxs were online in 2010. This audience is consistent with what Chávez (2015) describes as *new Latino*, an industry construction that defines the ideal Latinx as bilingual, culturally fluid, and technologically savvy.

This segment of the Latinx community has been courted widely by corporate marketers because they possess the requisite levels of social, cultural, and economic capital.

At the same time, the producers of the show have expressed an interest in developing a more inclusive strategy, which would include non-Latinxs. As Garsd stated in an interview with *USA Today*, "I'm just as thrilled when I hear from a retired Jewish lady in the Midwest who doesn't speak Spanish but says she loves us" (Alvarado 2013). Contreras has expressed a similar desire to reach non-Latinxs, stating: "I'm thinking about some, middle-aged white woman in a pickup truck, out in the middle of Colorado somewhere that happens upon us. I want to keep her engaged. Because she's going to be curious, she's going to be engaged, she's gonna want to know more. She's on the left-end of the dial, so she's looking for something. I want to help show her things."

Contreras's description of the curious NPR listener is a recurring trope that I encountered throughout my research for this book. It is a trope that goes back to the network's founding and is still leveraged widely in NPR's branding materials. In practice, however, creating a single show that can cross ethnic, class, and racial lines can be tricky. For Garsd, it is a matter of finding the right tone. This also means figuring out the degree to which cultural references would be left unexplained. As she stated, "are you trying to make a show for us or by us? Or are you trying to do an explainer? We also had a scope that not everything is for you. We had this debate. It was really healthy. You're having these conversations. Tone is super important. Are we sounding a little pandering?"

Contreras similarly expressed the challenge of trying to balance in-group and out-group dialogue. Here, he focuses on specific editorial practices:

> The example that I use is cumbia. When we first started doing the show, we were like, "now we're gonna play a cumbia from Argentina." And the NPR editor that was working with us when we first started making the show. She said, you've got to explain cumbia. And initially I was like, who doesn't know what cumbia is? But then she explained, you have to put the explanation comma in there.

And so we did, up to a certain point. Now we don't have
to.... I think long and hard about whether or not we use
the explanation comma.

Insight into how the producers of *Alt.Latino* negotiate their audience can be found in the name itself. When asked about how the
team landed on the name *Alt.Latino*, Contreras expressed that they
wanted it to be both clear and inclusive. "If we had called it something
like *Ear Candy*," Contreras told me, "nobody would know what the
show was about." Contreras also revealed that they originally considered the name *Pachanga* but then abandoned the idea because it
was considered "too Mexican" and therefore too limiting. They also
wanted a name that could be pronounced in both Spanish and English. In his words, "we wanted something like Univision, that had
that duality."

At the same time, both *Alt* (alternative) and *Latino* are defined
broadly, which means that locating the exact kind of music that would
be considered Latin Alternative can be elusive. At first glance, the
term *Latin Alternative* appears to be a negative designation, defined in
terms of what it is not. Consider how Garsd and Contreras (2010) described Latin Alternative during an interview: "We think the answer
lies in first describing what it is *not*: It's not just traditional genres of
Latin music (salsa, merengue, cumbia), nor is it American hip-hop,
indie or rock."

Like the term *indie*, *Alt* (alternative) is meant to signal an alternative to commercial radio, but further reflection reveals that it is
grafted onto preexisting market labels used by the music industry.
According to Contreras, they made a conscious decision to avoid focusing on Latino pop music, which they believed was already well
covered in the media marketplace. This broad definition of Latin Alternative enables the producers to transcend regional tastes, thereby
expanding *Alt.Latino*'s potential audience.

The producers describe this kind of a collective as a "mash up,"
but one that closely resembles what Ortiz (1994) describes as an "international popular culture," a polyglot made from fragments of different nations. Any given show might include some combination of
electronica, hip hop, Spanish punk, rock, and ranchera, but, in doing
so, *Alt.Latino* engages in the longstanding industry practice of col-

lapsing linguistically, ethnically, racially diverse groups under the rubric of *Latino* or *Hispanic*.

This kind of collapsing divests specific forms of protest of their power by disconnecting them from their local contexts. As Feld (2000) argues, commercial music may allow for spaces of global solidarity, but it also removes listeners from the actual social and political conditions that gave rise to the music in the first place. Consider "Songs that Shout Protest" (2019), a playlist generated by *Alt.Latino*'s producers and featured on NPR's website. As the title suggests, the playlist is meant to draw connections among artists who are calling attention to social issues, and it features musicians including Dellafuente (Spain), La Santa Cecilia (United States), and Ani Cordero (Puerto Rico). Within the framework of this playlist, the producers are making an implicit connection between La Santa Cecilia, who is addressing U.S. immigration policy, and Puerto Rican singer Ani Cordero, who is protesting government corruption. But it is also important to consider that the focus is less on the socio-historic conditions that led to each specific protest. Rather, it is the *sound* of protest that is paramount. As Feld (2000) argues, it is the detachability of sound from its specific cultural context that makes music more consumable for a transnational audience.

In interviews, both Garsd and Contreras have described this kind of pan-ethnicity as a form of border crossing, stating "borders and boundaries mean nothing to us. Latin Alternative is a little bit of everything from everywhere mixed into a completely new Latino soundscape" (Garsd and Contreras 2010). It becomes clear, however, that the producers are interested in a particular kind of border crossing. As Bauman (1996) points out, not everyone gets to cross borders on equal terms. There are those who cross borders seeking new experiences, but they travel on their own terms and always have a home to return to. Others, however, cross borders under dire circumstances and are often rejected by the residents of their host country.

Alt.Latino's specific cosmopolitanism sensibilities reflect the kind of "border crossing" done easily by members of a certain class of Latinx with legal status, monetary resources, and education. This is the kind of border crossing that is favored by large global media institutions. To appeal to their ideal listener, the show often assumes a "tourist gaze" (Urry 2002), in which the listener is guided, by knowledgeable insiders, through the Latinx experience. At times, the show itself acts as a form of ethnomusicology, educating listeners on Latino culture and history. This is embodied in the concept of the guest DJs, individuals who have the capacity to arbitrate between the Latinx community and white, educated, middle-class listeners. Often the guest DJs are Latinxs who would be recognized by white audiences, including Rita Moreno, Nobel Laureate Junot Díaz, and journalist Gustavo Aurellano, who was known for his column *Ask a Mexican*, a column that ran in the *OC Weekly*, in which white readers were encouraged to ask questions about Mexicans and Mexican culture.

Furthermore, Contreras also serves as a guest on other NPR programs, such as *Weekend Edition*, where he shares new discoveries in Latinx music for a primarily white, educated audience. By serving as a tour guide for Latinx music and culture, *Alt.Latino* extends a long-standing tradition in the radio industry, which is to treat Latinx music as a form of cultural tourism for mainstream audiences (Casillas 2014). For source material, the producers of the show turn to industry taste-makers who are showcased in a variety of venues, including South by Southwest (SXSW), the Latin Alternative Music Conference, Viva Latino (an international rock festival), and the Latin GRAMMYs.

Alt.Latino and Civic Engagement

Alt.Latino's capacity to promote counter-hegemonic discourses is largely predicated on the argument that the podcast serves as a space for Latinxs by Latinxs. As Contreras tells it, "[the show] is produced and edited by Latinos. That's why we got into the business, to tell our stories, our way." By laying claim to access, Contreras is signaling the larger promise of public radio, which is to provide a platform for voices that have been suppressed in the media marketplace.

Despite both the motivation and the access, however, *Alt.Latino*'s producers are still working for an institution that operates according to its own logics of practice and is beholden to its own economic pressures. This, in turn, has shaped the kind of oppositional work in which *Alt.Latino* can engage. It is clear that the producers of *Alt.Latino* have been able to negotiate the mechanisms of public radio to make clear and pointed stands against lawmakers, particularly on the issue of immigration. Furthermore, it is *Alt.Latino*'s status as a music-oriented show that allows its producers to address issues in ways that are not possible in NPR's news-oriented programming.

In doing so, it appears that the producers of *Alt.Latino* have been able to exploit NPR's commercial sensibilities to its advantage. After all, NPR made the decision to invest resources into *Alt.Latino* because they believed that it met a strategic need of the organization. U.S. Latinxs have been identified by NPR as a strategic priority because of their long-term importance to the economic well-being of the network. But the podcast also conducts other forms of labor on behalf of NPR. *Alt.Latino* generates revenue by cross-promoting other NPR assets and encouraging economic transactions. For example, the *Alt.Latino* microsite is hosted on the NPR website. Consequently, listeners are exposed to the same marketing devices. On the website, listeners can activate a "Donate Now" button, or purchase a NPR-branded T-shirt or water bottle by pressing "NPR shop." The songs featured on any particular podcast are accompanied by their respective videos, which are hosted on YouTube, where listeners are presented with more commercials. At the bottom of the homepage, listeners are directed to other commercial spaces where *Alt.Latino* can be accessed, including Spotify, Google Podcasts, NPR One, and Apple Podcasts.

For individual member stations that are considering carrying the program, there are the more immediate goals. In a weekly programming schedule that is finite, all programs must deliver. The producers of *Alt.Latino* recognize that its ability to thrive is predicated on its ability to bring the stations money. To encourage NPR member stations to air *Alt.Latino*, the show also exists as a thirty-minute radio version that is made available for free to station managers. By removing cost as a barrier, NPR executives had hoped to encourage its member stations to run more diversity programming. According to Contreras, "the stations haven't figured out a way how to monetize, in a significant way. You gotta pay the bills." Contreras was pragmatic about the ways in which the fate of *Alt.Latino* is linked to the short-term needs of stations: "[NPR has] outgrown its 70s hippie origins. It's changed dramatically. Its board of directors is primarily station managers. And what's good for the stations is not necessarily what's good for NPR and they'll do what's good for stations."

As Contreras points out, station managers figure prominently in NPR's board of directors (National Public Radio 2019), ensuring that the needs of stations are voiced. During our conversation, Contreras said he believes that the needs of the stations have superseded the interest of NPR, which has led to conservative decisions in their programming strategies: "I think some of these program directors are a little too conservative. I don't think they give their audience the benefit of the doubt. Which is why we're only on fifty stations and they're taking it for free. If we had to charge for it, I don't know if the stations would take it. Especially after 9-11, it's a lot of talk, a lot of information. Even *Jazz Profiles* is no longer there."

Contreras believes that *Alt.Latino* helps to expand the NPR brand by making it more inclusive. However, because the network does not make its audience data available, it is difficult to ascertain whether *Alt.Latino* has advanced NPR's goal of diversifying its audience. Furthermore, *Alt.Latino*'s articulation of the Latinx listener is different from some Spanish-language commercial stations, which have found success in focusing on working-class Latinxs with regionally specific tastes. Today, Mexican regional music radio continues to be one of the most listened-to formats (Nielsen 2018).

It is also difficult to ascertain whether the podcast helps deliver on another of NPR's original mandates, which is to engage listeners civically. Given its focus on acculturated Latinxs, it is important to consider the specific type of oppositional work in which *Alt.Latino* engages. When I asked Garsd if the many social issues have been converted into concrete civic action, she told me that she has received some inquiries on social media from listeners who are asking how they can help on a particular issue or where they can donate time and money. In some cases, it might be from someone who wants to start a nonprofit initiative. These examples are largely anecdotal, but there are no formal on-air efforts to increase public engagement around the issues that are being addressed.

Because there are no formal mechanisms for promoting civic engagement, it appears that *Alt.Latino* primarily serves as a form of consciousness-raising for listeners who already possess significant amounts of social and cultural capital rather than as a form of civic engagement for disenfranchised audiences. This kind of oppositional work can seem tepid when compared to the efforts of commercial Spanish-language radio. As I have previously discussed, there is a growing literature that has showcased the ways in which commercial Spanish-language radio stations have engaged in overt forms of civic engagement, which range from generating an overall sense of belonging (Casillas 2014) to providing information on naturalization and organizing voter drives (Baum 2006) and mobilizing direct forms of protest (Félix, González, and Ramirez 2008).

From this perspective, *Alt.Latino* also shows the clear limitations of using music to engage in oppositional messaging. When I asked if there was any evidence that *Alt.Latino*'s discussion of immigration has led to any kind of civic engagement, Contreras was hesitant. He was sure to distance himself from other forms of Latinx media, like Univision, which he believes pushes the boundaries that separate journalism and activism. According to Contreras, the role of *Alt.Latino* is to "inform, entertain, and educate," but not necessarily to directly engage listeners civically.

However, we also discussed the need to address Latinx issues in a climate in which anti-Latinx rhetoric has escalated across the country. Contreras admitted that he had felt an urgency to address immigration after the election of Donald Trump. We spoke about what it means to be a journalist working in the current climate. Contreras acknowledges that these attacks feel personal to him and agrees that journalism needs to take up that challenge. But he also acknowledges the struggle to balance journalistic objectivity with his personal feelings. "It's the same conversation that Latin journalists have right now," Contreras stated. "How do you cover this when you're personally attacked?"

During our conversation, we discussed how he would compare the kind of engagement he sees as appropriate for *Alt.Latino* to María Hinojosa's *Latino USA*, which was supported by NPR for over twenty-five years, but which has taken a more pointed position on Latinx-oriented issues. Contreras was reflective: "María's been very smart about establishing her independence. NPR distributes the show, but if they [*Latino USA*] started doing stuff that NPR felt uncomfortable with, then that relationship would end. At least with the radio stations . . . you could go right up to the line."

But, as Contreras makes clear, he is careful not to cross that line. He expressed a keen awareness that his personal feelings must be tempered while representing a network that is primarily a news organization. And, while he would like to see journalists take more of a stand, he recognizes that he is beholden to the values of the organization for whom he works. "I think he's [Trump] changed the rules to the point that journalism has to change," Contreras told me, "but my bosses don't think that. There's only so far I can go with Felix's theory."

REIMAGINING NPR IN A POST-WHITE AMERICA

I left *Talk of the Nation* in October of 1999. It wasn't until the spring of 2016, seventeen years later. Until then, there was no Latino program host at NPR. Come on, man. Seventeen years? Just through the course of things, there would have been progress. But it wasn't until Lourdes Garcia Navarro was doing a one day a week. That's pathetic. At that point, you're not even trying.

—RAY SUAREZ, (PERSONAL INTERVIEW)

A Separate Public

For twenty-one years, NPR produced *Talk of the Nation*, a news-oriented call-in show that focused on issues of national concern. There were episodes that centered on "The ABCs of Telecommunications" (1998), "The End of Civic Society" (1996), or "English as the Official US Language" (1996). Listeners were meant to play a meaningful role in the discussion by offering their viewpoint or asking questions of the host or the expert guests.

From 1993 to 1999, the show was hosted by Ray Suarez, a journalist of Puerto Rican heritage from Brooklyn, New York. At NPR and PBS, Suarez had focused on national issues, but he has also had an interest in how Latinxs figure into the public's imagination of the nation. In 2013, he wrote a book titled *Latino Americans: The 500 Year Legacy that Shaped a Nation* (2013), which is a socio-historic perspective of Latinxs in the United States. The book, and the PBS documentary series that accompanied it, were meant to disrupt dominant discourses that frame Latinxs as perpetual outsiders. In a reading at the Library of Congress (2014), Suarez distilled the essence of the book into three key points:

> One, we're not new. Our story is older than Plymouth Rock and Jamestown. Number two, a lot of us are *here*, because the U.S. was *there*. Invading Mexico, invading Cuba, seizing Puerto Rico, occupying Nicaragua, break-

ing Panama off of Colombia, and on and on and on. And three, once a group of people, any group of people, becomes a sixth of the whole, in this case more than fifty million out of 310 million, it's no longer a question of how "those people over there" are doing or whether or not they're making it. We are saying to the rest of America, we are you, you are us, our fates are intertwined.

Here, Suarez makes the essential point that, despite their long-standing presence in the United States, Latinxs continue to be framed as strangers in popular discourses. I asked Suarez how Latinxs' perpetual outsider status undermines their ability to be considered the *public* that public media is tasked with serving. Specifically, how has NPR conceptualized Latinx listeners as members of its listening public? "Fitfully. Inconsistently and unreliably," Suarez told me. But he does not think this issue is limited to NPR. "CPB, PBS, and NPR are all institutions of America, so they're held to the same cultural assumptions that America is taken to. But that means they wouldn't be removed from the American habit of mind." He went on: "This American apprenticeship never seems to have an expiration date. Like that politician in Wisconsin, who said people in meat packing are dying, but not regular people. These people who live in communities in large numbers. They work. They pay taxes. And they're not regular people?"

Suarez is referring to a statement made by Wisconsin Chief Justice Patience Roggensack, who attempted to explain the spread of the Coronavirus in Brown County, Wisconsin, by suggesting that the virus was limited to Latinxs working in the meatpacking industries. "That's where Brown County got the flare," Roggensack said. "It wasn't just the regular folks in Brown County" (Flynn 2020). Roggensack's statement echoed that of a Smithfield spokesperson, who seemed to blame Latinx workers for their own illnesses, arguing that "living circumstances in certain cultures are different than they are with your traditional American family," (Samaha and Baker 2020).

The characterization of Latinxs as being fundamentally different from "regular folks" and "traditional American families" reflects a broader presumption that there are unique differences that prohibit "Latinos" from being "Americans." These sensibilities will inevitably become evident in how we produce news. In 2019, NPR's Lourdes

Garcia Navarro wrote an article for the *Atlantic* (2019) in which she openly challenged the practices of dominant news organizations. Focusing on the shootings in El Paso that year, which left twenty-three dead, Garcia Navarro argued that news organizations failed the Latinx community by ignoring the growing animus toward them in the country. The racially motivated nature of the shootings was deftly explored in *Radio Ambulante*'s podcast "Una Ciudad en Dos" (Alarcón 2020), but this discussion was not occurring in mainstream news organizations, including NPR. According to Garcia Navarro:

> Latinos make up 18 percent of the population of the United States. Our roots here extend far back before the nation's founding. We have fought in every American war. Our food, our language, and our culture have shaped every aspect of American life, going back centuries. And yet the headlines in our largest papers and the cable-news chyrons omitted or downplayed the historic nature of the carnage in El Paso. Instead, they gave top billing to calls for unity by a president who has, for years, used angry rhetoric that dehumanizes and maligns Latinos.

By reinserting Latinxs into the story of America, Garcia-Navarro is reclaiming the right for Latinxs to be considered members of the listening public. Garcia-Navarro further argues that, while there has been a proliferation of Latinx-oriented media, which offer alternative perspectives, that media is walled off, while, in her words, "the pinnacles of elite journalism remain mostly white." Garcia-Navarro's assertion that the field of journalism is primarily a white public space echoes Amaya's (2013) argument that the media landscape has been shaped by intersecting capitalist and racial ideologies. Within a competitive marketplace, Latinx media has been equated with foreign media, but its relegation to the niche has limited Latinx media's ability to engage in national discourses.

The Commodification of Public Radio

There was a recurring trope that kept emerging during my interviews with public radio practitioners. Editors, producers, correspondents, and hosts consistently referred to people as "characters" in a "story." For example, when I asked Nadia Reiman, a producer for *Latino USA*, what makes a good episode, she told me, "it's very character driven, and very story driven. So, we're generally trying to tell a particular story," later adding, "we want the listener to be able to relate and identify with the characters, to feel what they feel."

Here's how *Radio Ambulante* talks about the people they cover:

> These are the characters that emerge from *Radio Ambulante* stories: a transgender Nicaraguan woman living with her Mexican wife in San Francisco's Mission District; a Peruvian stowaway telling his harrowing tale of coming to New York in 1959, hidden in the hold of a tanker ship; the Chilean soccer player who dared challenge the authority of General Pinochet; a young Argentine immigrant to North Carolina, trying to find his way through the racially charged environment of an American high school.

The practice of describing individuals as characters in a story is consistent with NPR's overall approach to journalism. In his book *NPR's Guide to Journalism and Production*, Jonathan Kern (2008) elaborates on this insight: "Almost every good story, from a children's fairy tale to an investigative piece for *The New York Times*, has a few basic ingredients. It has characters; it is set in some specific place; it has a clear beginning, middle and end; and there is some sort of tension that makes the listener or reader want to find out how things are resolved. Radio stories are no exception."

This guidance is also included on NPR's training site, which offers advice for journalists wishing to produce stories in the NPR style. According to NPR, journalists must find ways to build some empathy between the listener and the people featured in their stories. A good story is one that "features compelling character[s]. One way to get listeners involved it to make them care about someone in your story" (Macadam 2015).

This way of speaking about actual people with actual lived experiences has always struck me as curious. I teach in a school of journalism and communication, and it is how our faculty teaches aspiring journalists. I asked my colleague, who is an established journalist, why we use this term. He told me:

> I mean it's important for a writer to find someone who has a story, readers can connect to on an emotional or personal level. Someone readers can root for or against (more times than not, root for). Someone who has a "narrative arc"—we see that the person has done something/ achieved something, and as a reporter we go back and find out the events, the people, the issues that led up to this achievement.

And all this makes sense . . . from a certain point of view. After all, there is something fundamentally human about telling stories. Walt Fisher (1989) calls us *Homo Narrans*, storytelling man, meaning that stories have become a fundamental way in which we make sense of the world and our connection to others. This insight helps journalists to engage the listener by drawing them into the lifeworld of an individual.

But we must also acknowledge that there are certain kinds of ideological work at play when journalists present people's life experiences in this way. Real people are complicated, fractious, and inconsistent. Their lives often take circuitous paths rather than following a clear narrative arc. There are reversals and setbacks, and some events and experiences are never resolved. But, to create coherence out of incoherence, the writer must smooth out the rough edges. They must simplify and typify. Furthermore, they must present the "character" in a way that will appeal to the intended listener based on the presumption of shared similarity. Therefore, significant differences between the "character" and the listener must be minimized. *They* must be made to seem like *me*.

And then there are the economic implications to consider. Stories, in essence, serve as a form of commercial entertainment meant to attract listeners. Within a highly competitive media marketplace, in which listeners have an infinite number of options, engaging the listener is a matter of self-survival. If you capture enough listeners, then you have an audience that can then be sold to foundations and corporate underwriters. In short, the use of the term *character* reflects the form of public media we now have, in which news, comedy, and fiction all occupy their space as media products available to discerning consumers.

All this speaks to a much larger issue, which is that NPR, for all intents and purposes, has become a media company. And media companies will do what they will do. They will expand, they will seek efficiencies, they will centralize content, they will seek the economic resources that will allow them to grow. They will pursue audiences that serve their economic needs.

But, the more NPR becomes wedded to marketplace logic, the more it deviates from what it was intended to be in the first place. During our conversation, Suarez pointed out that NPR was meant to be distinct from commercial news organizations based on its core funding model and its mission. Today, those funding models have become uncomfortably similar. According to Suarez:

> What's the unique mission of public radio? That's less clear of an answer today than it was thirty years ago. Where every day here in Washington, you can hear Michael Barbaro on *The Daily*, a *New York Times* podcast, and also the *New Yorker Radio Hour*. They are unquestionably commercial. But they have a Venn diagram, where they share the same audiences with public radio. So no, the mission is not as clear.

Here, Suarez raises two essential points. First, there is increasing homogeneity, rather than diversity, within the media marketplace. Certainly, the growing presence of commercially produced programs that are slotted into the schedules of NPR member stations suggests that the boundaries that separate public and commercial are diminishing. Furthermore, when you look at the headlines of NPR, the *New York Times*, and the *Washington Post*, they are remarkably similar, supporting Bourdieu's (1990) point that news organizations are subject to economic pressures. Industry metrics such as audience ratings and fundraising efforts are essential to the growth of organizations, but they make NPR cautious and conformist, which contributes to simple reproduction of the field.

Second, these economic pressures will dictate the audience that NPR will pursue. NPR has actively cultivated an audience that is already engaging in civic discourses while excluding those most disenfranchised by the media marketplace. Suarez believes that NPR, like other commercial news organizations, has moved away from being a medium where licensees were targeting big chunks of the audience, what he called Middle of the Road stations. Now radio looks more like magazines, which are pursuing more narrowly defined segments of the audience. While this reflects the industry's overall shift from general to niche, Suarez believes that NPR has ceased to think of the listener as the public. "In commercial radio, the listener was the product," Suarez told me. "In non-commercial radio, the product was the product, and the listener was the public. Now public radio stations have rate cards, so what's the difference?"

I have made the argument that NPR has survived in this marketplace by targeting a narrow slice of the listening audience, which is affluent, educated, and primarily white. This argument is not new. Media scholars have deftly made the case that NPR has failed to achieve its mission of giving a voice to those excluded by the marketplace. In this book, my own interest has been on how economic factors have inhibited NPR's ability to enact its diversity mandate, with a particular focus on the Latinx community. Having a clear understanding of NPR's economic pressures is important to understanding how NPR imagines and serves the Latinx public. It also helps us better to evaluate the nature of the work being conducted by Latinx-produced programs like *Radio Ambulante, Latino USA*, and *Alt.Latino* that are produced in partnership with NPR.

There is no doubt that all three are excellent shows. They do good work. They tell important stories. In doing so, they raise the profiles of Latinx public figures and help advance the careers of Latinx journalists and producers. And they are highly regarded within their fields. Furthermore, they all have been able to accomplish something remarkable, which is simply to create a viable radio program that can ably compete for listeners in a crowded media marketplace. It also becomes clear that all three of these shows draw heavily from market rationality.

Several practitioners with whom I spoke, however, see promise for Latinxs in the marketplace model. When consumers choose some media products over others, they publicly define what they think is valuable, echoing Garcia-Canclini's (2001) argument that, through consumption, Latinxs may be able to claim a place within society and, eventually, rights. This was a perspective shared by Martina Castro, who helped launch *Radio Ambulante* and who is now president of Adonde Media, a podcast production company that has partnered with NPR to understand the Latino podcast listener. Castro is practical in her assessment that NPR needs to function as a business in order to survive in a highly competitive media marketplace. "It's about business sense," Castro told me. "I think public media is just as privy to those pressures, economic pressures, as any company. Advertisers want to advertise to affluent people who are going to buy their products."

Castro does not necessarily have an issue with NPR thinking about Latinxs as consumers. She just believes that NPR has failed to see their full market potential. Rather than focus on a narrow slice of the audience, she believes that NPR can follow Corporate America's lead and broaden the Hispanic market: "The problem is that they [NPR] don't realize that there are people in different spectrums of education and economic class that also buy things and are also worthy of being served where they're at, you know? This is the work, that mission of serving the whole public, could do a little more innovating. And not just continue to serve the same slice of America."

When describing his own thoughts on public radio, Bruce Theriault, who served as senior vice president of radio at the CPB, expressed a similar point of view: that it simply makes good business sense to ensure that Latinxs are incorporated into NPR's programming strategy. "From a business standpoint," Theriault told me, "it doesn't make sense to continue to narrow who you're talking to. It makes more sense to expand. So, I see it as a net gain, not a net loss." To Theriault, however, it is essential that NPR and its member stations not lose their core listeners in the process of pursuing Latinx listeners. He later added, "you're not trying to trade audience, you're trying to add audience. You're trying to reach a greater breadth."

This is an essential point. To add—rather than trade—audience, NPR must necessarily pursue a Latinx listener who is congruent with its existing audience profile. The Latinx cultural producers with whom I spoke were intuitively aware of this fact, and each, to varying degrees, drew upon industry research to deliver this audience. Whether it was the bicultural, tech-savvy Latinx identified by *Alt.Latino*, or *Radio Ambulante*'s Fusionista, these programs were selling a Latinx audience that possessed a comfortable amount of social, cultural, and economic capital.

Furthermore, each of the Latinx practitioners with whom I spoke was highly entrepreneurial. This is an essential trait if one wants to create a program that will thrive in a highly competitive media marketplace. From its outset, *Radio Ambulante* has been tremendously resourceful, beginning with the Kickstarter campaign that helped launched the show, to its more recent collaborations with ACAST, a global podcast company, and with tech company Jiveworld. Furthermore, the launch of the news podcast *El Hilo* shows the market potential for expanding into new markets that extend well beyond the nation-state.

This kind of business savvy is not limited to newcomers. Over several decades, *Latino USA* has evolved into a formidable media brand, led by María Hinojosa, who has become a successful brand in her own right. But adherence to market rationality has implicitly shaped the civic missions of these programs. Within a market framework, Latinx stories become objects available for public consumption. In turn, Latinx listeners perform their roles as citizen-consumers (Banet-Weiser 2007), a form of citizenship that recognizes economic potential rather than political action.

Audience size is one way of measuring success. Another is online activity, which, the participants believe, can serve as an indicator of civic discourse. "We had this amazing listener engagement," one journalist told me, describing the moment when she knew that a particular show was a success. Another journalist echoed this sentiment, stating "we knew the conversation was lit on fire on Facebook, in social media." Another journalist told me that positive feedback was an indicator that listeners were engaged, stating: "certainly, the feedback online is encouraging at least. When we get a lot of encouragement from people that listen."

But commenting, liking, and sharing all become forms of labor rather than discourses meant to prompt civic action. Consider how *Latino USA* talks about its listeners in its annual report to investors. The report includes a section titled "By the numbers," which provides insight into key metrics. We learn that, in 2017, Futuro's Facebook posts had over 42.5 million impressions, and they had a 48 percent increase in Facebook likes. They also had a 60 percent increase in Instagram followers, and their Twitter posts generated nearly 10.2 million impressions. These numbers are significant because they enable Futuro to measure and sell an audience that is active on social media to funders, station managers, and media partners. We also learn that, in 2017, Futuro Media had a 53.8 percent increase in institutional funders, which is another indicator of success. These kinds of metrics are not unique to *Latino USA*. They are simply playing the game in which radio audiences are delivered to prospective funders. These behavioral measures, however, do not account for the multiple ways in which listeners may engage civically, or the sense of belonging, which cannot easily be quantified.

Radio and the Civic Paradigm

Between 1935 and 1956, NBC's Blue Network carried a radio program called *America's Town Hall Meeting of the Air*, hosted by George V. Denny. Like *Talk of the Nation*, the hour-long program addressed civic issues of the day that were of importance to the American public. NBC paid the out-of-pocket expenses and provided the lecture hall, but the program was produced by the League of Political Education, a New York City-based group founded in 1894 to ensure forums where Americans of all ranks could be educated on important issues of the day. The league itself was funded through membership dues and donations.

Each show centered around a particular question that was posed by a member of the listening audience. "Should the President's Civil Rights Program be Adopted?", "Does America Need Compulsory Health Care?", or "Do We Have a Free Press?" Each of these questions would then be debated by a panel of experts. Take, for example, the question "Is Propaganda an Asset or a Liability in a Democracy?", the focus of a broadcast that aired in 1937. To interrogate this issue, the show brought together Anne O'Hare, journalist at the *New York Times*; Edward Bernays, known as the father of public relations; and Professor Harwood L. Childs, editor of Princeton's *Public Opinion Quarterly*. The hour-long format allowed ample time for opposing viewpoints and thoroughly laid-out arguments.

What was remarkable about the show was the active participation required from the listener. Broadcasts were always held in front of a live audience with up to one thousand people in attendance (Sterling and O'Dell 2011). Denny, speaking in a mid-Atlantic accent, would start the show by reading a letter from a listener. The audience could pose questions to the panelists, who answered their questions on air. While the actual debates were conducted in a lecture hall, usually in New York, smaller groups of listeners gathered in public libraries all over the country so that they could hold their own debates once the show ended (Lepore 2020). Transcripts of each broadcast were published by Columbia University Press as "Town Meeting: Bulletin of America's Town Meeting of the Air," and listeners could purchase copies for ten cents.

According to Goodman (2011), the program is indicative of the civic paradigm that was present in American commercial broadcasting during the 1930s. During this brief period in commercial radio history, broadcasters invested in producing an active, opinionated, and civically engaged listener. These efforts were likely a direct result of state intervention by the Federal Communications Commission, which required that broadcasters air low-profit public affairs. However, commercial networks began to see the strategic advantage in developing civic and educational programming. The networks realized that they could exploit these as a means of winning political favor as well as calm the demand for nonprofit broadcasting. But Goodman (2011) argues that this period was short-lived. These stations provided a model for cultivating engaged citizens, but they were ultimately subject to market forces and, therefore, doomed to fail.

As advocates continued to turn their attention to a separate, publicly funded system, there was a clear desire for public radio to assume the role of cultivating an active, engaged listener. It was also clear that stations could serve listeners in very direct ways by connecting them with health care services, serving as an educational resource, and engaging them civically. According to *The Hidden Medium: A Status Report on Educational Radio in the United States* (1967), the link between the listener and their civic institutions was meant to be direct: "Often the educational station provides the citizen with a vital link to his government on several levels; many stations serve as the only broadcast outlet for such public bodies as state legislatures, city councils and boards of education."

NPR was established with the expectation that the network would extend the mission of educational radio, only with the support of federal funding. When I spoke with Bill Siemering, he described how, in the early days of the network, there was some conversation about how NPR should engage the listener. "That was a subject of discussion at the board meeting. Were we going too far in saying that they would encourage active constructive participation, and so they can participate in affecting the process of change? So that's pretty active."

When pushed to describe the specific ways in which NPR empowers its listeners to have an active role in change, however, his vision for the network sounds much more conventional. "I think they don't go to the point of advocating action," Siemering told me. "But they give [listeners] the information they need to participate in the process of change." Siemering went on to describe the unique function of radio:

> I think that one of the things, just talking about radio as a medium, when you think of it as your radio bringing information to people. But its great strength, its unique strength is horizontal communication. The discussion, the debate, you know, hearing different points of view. That's overlooked too much in media studies, is that unique strength. No one else does that well. So in that way, you formulate your own ideas. And again, we're trusting the listener, assuming the listener is active. And when they have the information, they'll be even more active, perhaps.

In a way, Siemering is right. That is the strength of radio. Aspirations for public radio focused on the medium's ability to foster a strongly interactive relationship between the listener and broadcaster. But what Siemering is describing, and what NPR has become, sound more like a vertical form of communication, in which information is centralized and communication flows not across but downward from the few at the head of the hierarchy to the many at the bottom. Those audiences may choose to, or choose not to, act upon that information.

McCourt (1999) is particularly insightful in identifying three factors that impede NPR's mission to achieve its civic mandate. First, as a public good, NPR's benefits are intangible, yet public broadcasting has historically measured success quantitatively. Therefore, revenue generation and audience share become the primary markers of success. Second, despite its designation as a "public" resource, its organizational structure ensures that decisions are made centrally by organizations and professionals. Finally, NPR has come over time to prefer universality over diversity and locality.

NPR's aversion to direct forms of civic engagement was shared by almost all participants with whom I spoke. For example, NPR's former public editor, Edward Schumacher Matos, told me, "there's a limit to how much you can provoke civic engagement. But informing indirectly is helping that. You hope that it will help lead to civic engagement." This was echoed at the station level. A general manager with whom I spoke told me, "I don't know of any public radio stations that are involved in "get out the vote" efforts, because I think there's a thought that it would violate the editorial firewall. It's that we're just there for information, not to turn people out at the polls." When asked to describe the ways in which his stations promote civic engagement, another station manager described a passive approach. "Talking about civic issues," he told me, "or by bringing in experts who represent a wider range of voices."

It has been Spanish-language radio stations that have practiced this form of horizontal communication envisioned by Siemering. In many cases, these stations have created meaningful dialogue with their listeners, and, in doing so, have engaged them civically. As Casillas (2014) has pointed out, Spanish-language radio call-in shows often serve as forums in which DJs and listeners can discuss frustration with civic institutions and how to navigate them in very specific ways. Stations, in turn, will create a space for lawyers, health professionals, and advocates to provide information that listeners can use to negotiate their civic lives. However, these efforts are not developed in spite of market forces, but because of them. These stations may be driven by a social mission, but this mission is congruent with their economic realities. Many Spanish-language stations target Spanish-speaking, working-class Latinx listeners, and efforts must be made to ensure that the listener is well served.

I found that the Latinx-oriented programs that are affiliated with NPR do not share the same sensibilities about civic engagement as their Spanish-language counterparts. Instead, they reflect NPR's practice of cultivating the passive consumption of news. Any evidence that the stories produced by NPR's practitioners have led to any sort of political change was purely anecdotal. Nor did the public radio practitioners with whom I spoke with express any real interest in promoting this sort of engagement. Almost universally, the practitioners with whom I spoke were insistent that their role was to tell important Latinx stories, but not necessarily to engage disenfranchised Latinxs in direct forms of civic action.

This perspective is an extension of NPR's overall notion of passive, disinterested journalism, which sees direct forms of civic engagement as forms of advocacy. NPR has, over time, become more orthodox in its unflinching belief in objectivity, but this practice was proving to be inadequate given the racial climate that existed during the time in which this book was written. The racial animosity demonstrated by the Trump administration had caused journalists to ask important questions about whether the journalistic notion of objectivity served the Latinx community. I spoke with former NPR public editor, Elizabeth Jensen, who told me that NPR journalists were at the time attempting to negotiate these issues:

> It's a challenge in the newsroom. Forget about casting it as a Latino issue, it's a challenge in the newsroom if you're gay and the president or vice president is anti-gay. It's a challenge in the newsroom if you're a woman and some of the news is and some of the initiatives are considered anti-women. . . . NPR does sort of take that neutral role and journalists do think that's appropriate. But I will say that people in the newsroom are talking about this. How you can't just be objective when you're a targeted person?

Attempts to dismantle Deferred Action for Childhood Arrivals (DACA), the Trump administration's child separation policy, opposition to the Equality Act, and efforts to roll back pro-choice legislation have all caused some NPR journalists to reflect on their practices. But NPR has struggled to reconcile these tensions. Instead, it remains steadfast in its position that journalists have a duty to simply report the news, free of any personal bias.

But this pretense of objectivity can sometimes distort, rather than reveal. For much of the Trump presidency, for example, the network was disinclined to call the president's comments racist, arguing that, to label a comment racist, one would have to know the intention, something a journalist could never know. It was not until 2019, three years into the Trump presidency, that NPR addressed the issue in an essay titled "'Racist,' Not 'Racially Charged': NPR's Thinking on Labeling the President's Tweets" (2019d). In it, Jensen argues that there has been enough evidence to suggest that Trump's policies and comments reflect racial animus.

By that time, however, NPR was only acknowledging what had already become obvious, that Trump was engaging in racist practices. When I spoke with Theriault, we discussed the notion of journalistic objectivity during a time in which the administration was actively challenging the core assumptions of journalism. Theriault told me that this was a point of discussion amongst public media practitioners: "Objectivity is kind of one of the core beliefs in journalism. Objectivity, truth, fairness, some balance, but not taking a point of view. And not coming up with solutions. But now there's a movement to say that objectivity is bullshit. Nobody's objective. You can't not have a point of view. Things are just flat out wrong. You can call that out."

NPR's unquestioning belief in journalistic objectivity refuses to acknowledge what social anthropologists have already begun to confront: that professionals who broker in "truth" are inclined to cling to a myth of detachment, which obscures their role in perpetuating social inequalities. Through our lived experience, we gain particular kinds of insight and develop a particular set of biases. As we become inculcated into our professions, we acquire yet another set of biases (Rosaldo 1993), which predetermines which experiences will be collected and how they will be reported.

NPR in a Post-white America

It has become evident that NPR's practices, and its membership, have becoming increasingly at odds with a country that is becoming more ethnically, racially, and linguistically diverse. Furthermore, as a national news organization, NPR is confronting the reality of a post-white America, in which Latinxs have a significant presence. Significant change, of course, would require a massive restructuring of NPR's infrastructure, something that large, heavy organizations are not inclined to do. The question then becomes: What are the possibilities, within the existing market framework, to ensure that Latinx listeners are adequately served by NPR? During my research, I encountered a number of tactics meant to address these issues. None of these recommendations are necessarily new, and some date back to NPR's early days. These recommendations came up, in various forms, during my review of industry documents and in my conversations with practitioners. The basic blueprint for more inclusion tends to fall into five categories. Each of these categories is interconnected, but can be isolated for analytical purposes:

Build a Pipeline of Talent

During my research, I interviewed a junior reporter who worked in a large metropolitan market. The journalist attributes her career at an NPR member station to a couple of initiatives, including the Raúl Ramírez Diversity Journalism Fund, which was established for San Francisco State University students. She also participated in Next-GenRadio, a fellowship sponsored by NPR, which provides professional training opportunities for college students, recent graduates, and early career professionals.

She is exactly the kind of journalist who, with enough opportunity, could transform the newsroom. She feels an obligation to cover communities of color and feels like she is empowered to do this by her station managers. She wants mentorship, but she reported that she has not found anyone to fill this role. "I always seek mentorship, wherever I go," she told me. "And I generally try to look to for people who maybe have the same experience as me. For me, that racial connection is very important. And that not's not as easy to find at [my station]. I haven't been able to find that mentor."

The journalist noted that she is often the only person of color in the room, and she described some of the microaggressions she experiences in the newsroom. But, overall, she believes in the importance of local public radio, she believes in the need to have a have a defined beat on race and ethnicity, and she believes there should be more women of color in leadership positions.

Her presence at a thriving NPR member station is remarkable. A basic reality is that there is a lack of newsroom diversity at NPR. This was an issue that was identified as far back as 1977, when the CPB's

Task Force on Minorities in Public Broadcasting found that there was only one "minority" employee involved in NPR's programming decisions (Corporation for Public Broadcasting 1978, xiii). There has been little progress since then. By 2019, Latinxs were still severely underrepresented at only 6.24 percent of NPR staff (Jensen 2019a).

An ongoing solution to this issue has been to build a pipeline of talent through fellowships and training programs, like the NextGen program. Similar efforts have been attempted over the years, but they have yet to yield any long-term change. As Theriault told me, "there have always been efforts at the Corporation for Public Broadcasting to increase diversity. There were women and minority training grants, but they were marginally successful."

One of the obstacles to inclusion is the issue of career advancement. There is a need to provide a clear pathway toward upward mobility, which can be a challenge given the limited number of leadership positions in public radio. This journalist, for example, will undoubtedly face a logjam at the managerial level. This is an issue that came up in several interviews that I conducted. As one news director told me: "This is an interesting challenge. Low turnover equals less diversity. In public media, you have a lot of long timers who stay for the state benefits or because of the mission. This impedes the entry of a lot of younger, more diverse folks."

This was a sentiment that was shared by the CPB's Bruce Theriault. During our conversation, we spoke about a station manager we both knew, who had recently retired:

> The workforce is smaller than a lot of industries. The turnover is less. [Said general manager] just retired, but he's been taking up a managerial spot for the last thirty years, at one station or another. And within, there isn't the structure for people to move up. There are big jumps between levels. And there just aren't as many opportunities. That station is one of those places where they're born there. And they die there.

To address the logjam at middle-manager and senior-executive levels, NPR and its member stations must either incentivize turnover or create new positions. But a massive shake-up at the senior level seems unlikely. It is also unlikely that new positions will be created, given the limited financial resources available to NPR and its members stations. It is more likely that new audio platforms, both commercial and noncommercial, will provide more opportunities and create new pathways in and out of public radio. As more outlets emerge, however, there will be greater competition for strong talent, which raises the issue of retention. Efforts must be made, at both the national and member-station level, to ensure that Latinx practitioners feel that their work is valued and that they are in the position to make meaningful contributions to the newsroom.

Cultivate New Sensibilities

For a period of time in the early nineties. María Martin served as Latino affairs editor at NPR. In this role, Martin's job was to serve as a sort of expert on the Latinx community with the goal of influencing story selection and editorial choices. However, Martin has maintained that the journalists and news directors with whom she worked never truly embraced this role. "I don't think they [NPR] got people who were different," Martin told me. "I was the member of something called the NPR affirmative action task force. And we did a survey of people of color who had left NPR. And almost to a person, everyone spoke about their difficult experience there. And the fact that it was white liberals, who thought they were very tolerant, and knew everything, they were the ones who gave them the problems."

Here, Martin describes climate issues at NPR, which she believes have historically marginalized practitioners of color. My own interviews with Latinx practitioners reveal varying levels of disenfranchisement. This is not necessarily surprising. As Bourdieu (1990) argues, when newcomers enter a field, they bring with them professional norms and dispositions that clash with the prevailing norms of production and the expectations of the field. This challenge is exacerbated by the fact that fields replicate themselves by hiring individuals who fit seamlessly within those fields. Overall, I found that NPR and its member stations tend to hire and promote talent that was groomed in public radio, which tends to have a conservative effect.

But conservative practices are ineffective during moments of transformation. Rather than simply inculcate journalists of color so that they adopt the practices of the field, it is equally important to challenge those norms by cultivating new sensibilities. According to *Reporting on Race in the 21st Century* (Aspen Institute 2015), this process should begin in journalism schools, where aspiring journalists are first taught the normative ideals and practices of their field. But this form of professional training should also be extended to established practitioners. Learning how to challenge existing assumptions and having a deeper understanding of the federal, state, and local laws that impact communities of color may help current public radio practitioners better understand how racial dynamics shape social policy.

Integrate Diverse Sources

After listening to NPR's news programs produced over the years, two things become clear. First, the segments on NPR's magazine programs have become much shorter. Second, there is a heavier reliance on expert sources. Former NPR public editor Elizabeth Jensen told me that she believes that tighter show clocks have pushed out a greater range of voices. At the same time, there has also been a push to go live, which means producers are reluctant to put on participants whom they believe may not perform well for live radio. Journalists, producers, and news directors prefer to work with experts, often government officials or academic researchers, who can provide context and analysis for their stories. Because they are underrepresented in these spaces, Latinxs do not benefit from this model.

When I raised these issues with one senior producer, he told me the decision to exclude Latinxs goes beyond practical considerations. "I disagree. When we had a longer format, we still didn't include Latinos on the show." According to the participant, a greater inclusion of Latinx perspectives and voices means NPR professionals being comfortable with Latinx voices. As he told me, "some people don't have a comfort level with people who don't look like them or sound different." The *Journalism in the 21st Century* report provides some direction on this issue. The report urges journalists to make a greater effort to gather the opinions of those who are directly impacted by the events, policy, or legislation on which they are reporting, which includes community leaders, activists, and residents. These sources are essential to providing important insight and historical context on any given issue. But linguistic proficiency can be an important factor. The ability to engage Spanish-speaking sources requires journalists who have the requisite linguistic proficiencies to engage these sources in meaningful ways.

Radio Ambulante has been particularly effective at producing stories from the point of view of those most directly impacted by particular policies. Their capacity to do so is possible because of the workforce they have. With a team made up almost exclusively of Spanish-speaking Latinxs and Latin Americans, they have the ability to seek out stories across the Americas, interview their sources in Spanish, and capably tell their stories.

The report also encourages newsrooms to navigate the complexities of race by consulting with nonprofit news organizations, researchers, and advocates during the reporting process. These experts can provide unique insights on research, policy issues, and events that impact their communities. This was something that *Latino USA* did, which led to *The Strange Death of José de Jesús*. When Hinojosa travelled to Arizona on a business trip, the producers at *Latino USA* reached out to Puente, a local community organization in Arizona. It was through this connection that they first learned about the case.

For those newsrooms that do not have established contacts, there are resources. In 2003, American Public Media (APM) launched a resource for public radio journalists called the Public Insight Network (PIN), which is a collection of web-based tools where participating stations can tap into a network of sources who share their insights, life experiences, and firsthand knowledge. As APM states, "source contributions are captured in a robust, secure, and searchable database which allows journalists to pose questions to sources based on demographics, geography, expertise or interests" (American Public Media, n.d.). According to APM, these kinds of insights can be used early in the reporting process to test hunches, gain unique insights, and explore possible angles.

Community Engagement

As institutions that are meant to reflect their communities, NPR member stations are encouraged to engage listeners in ways that extend beyond simple broadcasting of content. To serve this role, there is the potential to sponsor town halls, information sessions, and forums on community concerns. These forums not only give back to the community, but they also can provide news organizations with insights into emerging issues.

The *Journalism in the 21st Century* report encourages journalists to actively solicit audience feedback, to be responsive to that feedback, and to prioritize increasing coverage of issues of key importance to underserved communities. It also highlights possibilities with social media platforms, through which news organizations can create online forums for audiences to communicate directly with reporters and editors. The report also suggests that news organizations encourage citizen journalism by creating a space to air or publish contributions from the public.

Radio Ambulante has been effective at creating community with its transnational audience of listeners and, in doing so, building financial support for their work. As the organization has expanded, *Radio Ambulante* hired Jorge Caraballo as its growth editor. "I consider any audio project to be a conversation with a community," Caraballo said during an interview with *Storybench* (Aguirre 2019). In an effort to build these relationships, the team has developed "Clubes de Escucha," listening clubs gathered across the Americas, which allow members to engage *Radio Ambulante*'s podcasts in meaningful ways. This approach to audience engagement is reminiscent of *America's Town-Hall Meeting of the Air*, in which radio programs were designed to prompt active discussion among listeners around issues of shared concern. According to Caraballo, these spaces allow communities of *Radio Ambulante* listeners not only to dig more deeply into stories produced by *Radio Ambulante*, but to build stronger connections among themselves (de Assis 2019).

Invest in New Content

As early as 1977, the CPB's Task Force for Change found that programming, by and for minorities, was "seriously deficient" (1978). Yet, getting NPR to commit to Latinx-oriented programming has been a challenge over the years. The network's shift in strategy, which began almost forty years ago, has undermined any effort to create specialized programming. Whether it is labeled as a "unified appeal" approach or a "strategy of transcendence," the network has remained beholden to a programming strategy that narrowly focuses its core listener. According to Theriault:

> I think if you go back to some of the original, early work in public radio, the research, you'll find that public radio directors were discouraged from having things that didn't sound like public radio. And I kept challenging that because the definition was self-perpetuating. Well, we basically designed ourselves for college-educated white people. And then we said, this is how you get more of them. You don't do anything differently. Nobody said in the research, don't hire anyone with an accent, but they basically said there's a sound and if you deviate from that, you do so at the expense of your audience.

As Theriault suggests, by defining itself as a white public space, NPR has, in essence, pushed Latinx voices to the periphery. Consider *StoryCorps*, a project that collects oral histories. The effort benefited from its close relationship with NPR, which has featured edited versions of these histories on *Morning Edition*. The response from public radio listeners, a predominantly white and affluent audience, was strong, and they took up available time slots at *StoryCorps* booths. In the process, they shut out minorities who had never heard of *StoryCorps* and its efforts to collect oral histories (Goldsmith 2016).

To ensure that the stories of Latinxs could be heard, *Story Corps* responded by creating *Historias*, a dedicated space for Latinxs. According to *StoryCorps* (2020), the goal of *Historias* has been to record "the diverse stories and life experiences of Latina/Latino people in the United States. Sharing these stories celebrates our history, honors our heritage, and captures the true spirit of our community. It will also ensure that the voices of Latina/Latino people will be preserved and remembered for generations to come." These oral histories are archived at the American Folklife Center at the Library of Congress and in a special collection at The Nettie Lee Benson Latin American Collection at the University of Texas at Austin. But, because these stories are recorded in Spanish, they will not be heard on NPR because of the network's strict adherence to English monolingualism.

Podcasting may provide a pathway for Latinx-oriented content. While podcasts still remain overwhelmingly white (Friess 2017), NPR sees the potential for podcasting to reach diverse audiences (Powell 2015). The network was one of the sponsors of the *Latino Podcast Listener Report*, a study conducted by Edison Research (2020), which found that U.S. Latinxs were an important growth market in the podcasting space. Despite this potential, NPR has been slow to provide the institutional support for Latinx-oriented podcasts. This lack of support was one of the factors that led Hinojosa to partner with PRX as *Latino USA*'s official distributor. Today, *Alt.Latino* and *Radio Ambulante* are the only Latinx-oriented podcasts distributed by NPR.

Again, this speaks to the fundamental dilemma faced by NPR managers, which is how to reach new audiences with dedicated programming without isolating their legacy audience. One participant characterized this as a zero-sum proposition, in which station managers must choose between "stand-alone," specific programs that target specific audiences, and a "mainstream" approach, in which cultural fluency is integrated throughout all programming. According to the station manager:

> With NPR, I think there would be kind of a consensus that NPR has not been particularly successful in integration. It hasn't been. It remains primarily a white intelligentsia. A chattering class phenomenon. It's a white audience, historically. And it's been something that NPR has tried to crack. And the dynamic is this. Do you have these stand-alone programs, or do you mainstream the voices?

To illustrate the differences between stand-alone and mainstream, he invoked two examples: *Latino USA* and KPCC, an NPR member station in Los Angeles: "*Latino USA* is a vestige, perhaps successful, of being stand-alone . . . the contrasting vision is KPCC, which is to mainstream. It is predicated on [the belief] that if we have Latino voices hosting and setting the editorial direction for the program, then that would be successful in attracting Latino ears. And I think that it's widely believed that that's true."

The idea that member stations should adapt their talent, operations, and programming to better reflect the country sounds intuitively true, but it is not something that is practiced widely. When I asked the participant if this was a practice he followed at his own station, he demurred. After all, investing heavily in the recruitment of Latinx talent, hiring the requisite news editors and producers, cultivating new networks, and revisiting programming all take significant economic and temporal resources. As he told me, "finding a journalist to infiltrate an untapped community is a luxury that we can't afford. *The New York Times* can do this, but they're resource rich."

Re-evaluating the KPCC Success Story

These five categories of recommendations represent relatively conservative changes that are meant to be implemented within the existing public radio framework. However, their impact will be limited without larger transformation change. This became evident with KPCC in Los Angeles, which came up multiple times during my research as an example of how the CPB can cooperate with NPR member stations so that they better represent the demographic realities of their communities. *Current*, the industry's de facto trade publication, described KPCC as a "model for other stations seeking to diversify their audiences" (Falk 2015b). For stations hoping to duplicate KPCC's success, the LPRC (2015) analyzed the station in a case study as a blueprint for engaging Latinx listeners.

KPCC's success within public media has been attributed to following this basic blueprint. They hired strategically. With a grant from One Nation Media Project, the station invested in the hiring of A. Martínez, who went on to headline a marquee show, and has since stepped into the role of host of NPR's Morning Edition in Washington. In doing so, they went outside the public radio pipeline and hired him from sports radio, which ultimately worked in their favor. As mentioned in the LPRC report (2015), they had intended to cultivate a new sound that borrowed more from commercial radio. They also invested in the creation of *Take Two* after Madeline Brand's exit from the show.

The station also focused on changing its organizational culture by changing their hiring practices and investing in professional training. According to the LPRC report (2015), when hiring new talent, KPCC "focused on people who are comfortable with differences and realize that differences within an organization make it stronger" (Latino Public Radio Consortium 2015, 11). For people already on staff who did not share these same sensibilities, the station conducted numerous training and team-building exercises (Latino Public Radio Consortium 2015).

The true strength was KPCC's investment in community engagement. When it launched in Los Angeles, KPCC hired Edgar Aguirre to serve as managing director of external relations and strategic initiatives. Aguirre devoted much of his time to building connections among individuals and organizations in the Latinx, Asian American, and Black communities in the greater Los Angeles area. Aguirre was also able to negotiate partnerships with local foundations, businesses, alumni associations, legal centers, community organizations, and media organizations. KPCC also conducted a number of focus groups and meetings with civic leaders, with the goal of helping them to assess which issues and stories were important to members of the local community and which stories were not being told or were being told incorrectly.

By some available measures, these moves appear to have been a success. From Spring 2009 to Spring 2014, KPCC's total audience increased by 27 percent, while the number of Latino listeners increased by 96 percent. Their morning show, *Take Two*, is the most listened-to locally produced program on the station and has substantial Latino listenership, largely attributed to their host A. Martínez.

All of these efforts build a solid case that KPCC has solved NPR's Latinx issue. Upon closer examination, however, KPCC does not look dissimilar to other NPR member stations in large metropolitan areas. The station's local program, *Take Two*, does constitute original programming, and its headliner A. Martínez was an important hire. But a glance at KPCC's weekly scheduled shows includes a typical mix of commercial and noncommercial programs. *Latino USA* airs on Thursday nights, but there is also a steady diet of the BBC World Service, the *New Yorker Radio Hour*, and *Marketplace*. And, of course, *Morning Edition* and *All Things Considered*. Despite its tagline "We Speak Angeleno," none of KPCC's programs are produced in Spanish, which is widely spoken in the area. As a result, these changes feel like differences in degrees, not differences in kind, especially when you consider what the project was originally meant to be.

I spoke with Frank Cruz, former chairman of the CPB, who served two terms. Cruz co-founded Telemundo and went on to have a celebrated career in both English- and Spanish-language radio. His broad experience is likely the reason that President Clinton appointed him to the CPB. According to Cruz, it was his priority to ensure that public media support was channeled into efforts that would ensure that diverse communities were represented, stating: "The thing, of course, that I always advocated for. When Congress allocates funds to the CBP, which then turns around then sends programming dollar to hundreds of radio stations. That money goes to all Americans. It's not just white Americans. It's for all Americans."

During our conversation, Cruz told me he was able to have some impact on public television. In his autobiography, Cruz (2019) proudly discusses his role in developing some Latinx-oriented programs, including the drama series *American Family* and the animated children's program *Maya and Miguel*. But he acknowledged to me that he had less success with NPR. "I wish I could tell you that I did a comparable thing at public radio," he admitted.

In some ways, this is an issue of medium. Theriault told me that television is program-driven, meaning that resources can be channeled into specific programs to diversify the schedule. By contrast, Theriault argues that radio stations are destinations. According to Theriault: "I think it's actually the nature of radio versus television. People watch television programs, but they listen to radio stations. Stations have their overall appeal, but as they stray from that appeal, with specialized programming, they actually become a different a station, and it hurts them overall. And it's not very successful usually."

To have a significant impact on radio, the CPB felt that they needed to secure a station that could serve the Latinx community. Both he and Cruz saw an opportunity to make a difference in Los Angeles, a city with a significant Latinx population. The result was an initiative called Los Angeles Public Media (LAPM), an effort within the CPB to produce Latinx-oriented content in Spanish and English. I asked Cruz about his original vision for the project. He told me:

> What we were trying to do? I would always say that there are such high demographic numbers in the LA basin. We could very easily do well on the public radio and do it in Spanish language. They [Latinxs] can benefit from news on immigration. They can benefit from news on politics

and education for their kids. And since they're tax payers themselves, it's good for them to learn about what their government is doing.

The initiative began as a production center based out of NPR West that could produce content, in Spanish and English, for various public media outlets, but the team at the CPB had larger ambitions. The goal was for CPB either to secure a frequency or to partner with an existing station that would serve working-class Latinxs, who were currently not being engaged in civic discourses.

During our conversation, Bill Davis, president of KPCC, told me that the Spanish-speaking Latinx listener in Los Angeles is "over-served." But Davis was referring primarily to music-oriented stations. However, Cruz believed that a news-oriented, public radio station for working-class Latinxs would offer something unique in the marketplace:

> There are the Mexican workers in the plants, the garment industry, and the hotel workers. That's who we need to reach. Yes, they listen to commercial radio in Spanish, but that's music. What public broadcasting can do in Los Angeles. What I wanted them to address was to address those Hispanics that speak Spanish and can become better citizens in our country if they had the information. What Radio Bilingüe did with rural Latinos, we needed one for the urban Latinos.

Here, Cruz is reaffirming public radio's obligation to serve the civic needs of its listener, and he looked to Radio Bilingüe, which has been successful at fostering civic engagement among Spanish-dominant, working-class Latinxs, as his model. This articulation of the audience seems much closer to the audience that was identified by NPR's early architects. The station also broadcasts multilingually, in Spanish, Mixtec, and Triqui. Among the original ideas was the idea to help Fresno-based Radio Bilingüe have a presence in Los Angeles. To put this idea in motion, Cruz convened a meeting with various radio producers, which included the CPB, Radio Bilingüe's Hugo Morales, and, later, KPCC's Bill Davis.

At this point in the process, the team invested in audience research in an effort to ascertain the kind of listener they wanted to serve. Sponsored by KPCC, the research pointed to English-speaking Latinxs as a likely audience, which became a point of contention with Hugo Morales, the founder of Radio Bilingüe. "One of the early decisions," Theriault said, "was that it would be mostly in English. There would be some Spanglish. Hugo [Morales] came to that very reluctantly." During our conversation, Cruz also described the role of market research:

> One of the barriers was how do you reach that Hispanic audience. Do you do it English? Do you do it in Spanish? Do you do it bilingually? How do you reach them? They [KPCC] were doing a lot of studies. And they broke down the Hispanic audience in LA, the greater Southern California audience. How many of them speak Spanish only, how many English? I think what happened is that KPCC model. That perspective that won out. They said, it best to concentrate on Hispanics that speak English, because that's the Hispanic segment that is growing.

As Cruz suggests, attempts to overhaul a station that would serve working-class Latinxs failed, and the CPB withdrew funding. As a result, the LAPM project became defunct. Theriault expressed some disappointment that the project never took off in the way that he and Cruz had imagined: "The LA, Latino Public Radio project was the effort to try to make something happen in Los Angeles. To really establish a public radio and Latino presence that was owned and run by rather than just reach out and serve. We put millions into it, but I was overruled from putting more money into it, because it didn't look like it was going anywhere."

Once the LAPM project failed, the CPB supported KPCC's request to provide initial funding for a marquee show that would serve the Latinx community, what ultimately became *Take Two*. According to Theriault, "when it came to supporting KPCC, and supporting Bill Davis and A. Martínez, we really pushed it and provided a lot of money for it." Still, Theriault admitted it was not quite the vision he and Cruz had for a public radio station in Los Angeles. "Sometimes maybe the funder wants it more than the reality on the ground will dictate," he told me. "So, I was accused of getting Stockholm syndrome. Losing my perspective on what was possible and sinking more money in."

When we consider KPCC in its broader context, it becomes apparent that, despite a genuine desire to serve the Latinx listener, there is a gravitational pull toward sameness. Industry practices all but ensure that a Latinx-focused public radio station, in a heavily Latinx media market, will ultimately look and sound like other public radio stations in large metropolitan markets. A reluctance to broadcast in Spanish contributes to this process. NPR's unflinching commitment to English monolingualism ensures that many working-class, Spanish-dominant Latinxs listeners are shut out. Consequently, the boundaries that separate Latinx public media from the public media remain firmly entrenched.

This perceived binary of a national media system and a Latinx media system manifested itself in several of the interviews I conducted. Very early on in my research, one station manager advised me to reconsider my approach to the topic. "I'm still not completely sold on your framing," the station manager told me. "I would frame it as how public radio has attempted to diversify and reach listeners. And so there have been many failed experiments, but there have been remarkable episodes of success. And I think that's an untold story."

The untold story that the station manager preferred, however, was one of two separate public radio systems: Radio Bilingüe and NPR. He touted the important work done by Radio Bilingüe's Hugo Morales on behalf of Latinx agricultural workers, but, when it came to NPR, the station manager had less to report. He mentioned the occasional Latinx correspondent or Latinx-oriented program. Or he would refer to a specific station in a specific market that was doing good work. All these examples are true, but they did not add up to a coherent vision for how NPR integrates the Latinx listener in any consistent and meaningful way.

The inability to see Latinxs as the American public represents a failure of imagination that prohibits NPR from fulfilling its role of serving the public. After all, two separate systems do not reflect the social realities of Latinxs, who are often negotiating multiple public spaces. We must also attend to the fact that NPR does not operate in a vacuum. It is part of a larger media ecosystem that is itself embedded within a culture that continues to see U.S. Latinxs as a separate public. Reconstituting NPR so it is designed to better reflect a more diverse set of listeners, with a particular focus on those currently left out of civic discourses, would require transformative, rather than conservative, change. In a highly competitive marketplace, however, media executives are not inclined to take chances, despite the changing demographics. One participant with whom I spoke described it as a matter of risk. It is better to manage NPR into a "gradual decline" than it would be to make transformational decisions, which carry the risk of taking NPR into a "precipitous decline."

There is, of course, a cost for doing nothing. Now that the boundaries between commercial and public radio have been diminished, new players are entering the audio space with a similar mission and a similar storytelling approach. NPR is now at risk of losing its point of differentiation. But, as it clings to the business model that has served it well in the past, younger, more diverse viewers will look elsewhere for content they find relevant. Consequently, NPR risks being one of those brands who have failed to adapt to a changing marketplace. More tragic, however, is that it will have walked away from its mandate of serving the public interest. If NPR continues to maintain the status quo, it will become more and more out of sync with a country that has evolved. By 2045, the United States will become minority white, and, by that time, Latinxs account for almost a quarter of the population (Frey 2018). Unless it adapts, NPR will simply be a white public space in a post-white America.

REFERENCES

Adorno, Theodor W. 1945. "A Social Critique of Radio Music." *Kenyon Review* 18 (3/4): 229–35.

Aguirre, María Silvia. 2019. "Lessons on Building a Podcast's Community, Both Online and Offline, from Radio Ambulante." *Storybench*, October 10. https://www.storybench .org/lessons-on-building-a-podcasts-community-both-online-and-offline-from-radio -ambulante/.

Alarcón, Daniel. 2006. *War by Candlelight.* New York: Harper Collins.

Alarcón, Daniel. 2007. *Lost City Radio.* New York: Harper Collins.

Alarcón, Daniel. 2012. "Los Polizones," May 14, in *Radio Ambulante,* produced by Nancy López, podcast audio, 12:21, https://radioambulante.org/transcripcion/transcripcion -los-polizones.

Alarcón, Daniel. 2016. "¿Qué le Pasó a José de Jesús?" July 22, in *Radio Ambulante,* produced by Fernanda Echávarri, podcast audio, 32:29, https://radioambulante.org/en/audio-en /what-happened-to-jose-de-jesus.

Alarcón, Daniel. 2017a. "Storytime at the Azteca Boxing Club." In *Reality Radio: Telling True Stories in Sound,* 2nd ed., edited by John Biewen and Alexa Dilworth, 90–96. Chapel Hill: University of North Carolina Press.

Alarcon, Daniel. 2017b. "Cuando La Habana era Friki," January 24, in *Radio Ambulante,* produced by Luis Trelles, podcast audio, 23:01, https://radioambulante.org/en/audio -en/when-havana-was-friki-2.

Alarcón, Daniel. 2017c. "El Fotógrafo," November 7, in *Radio Ambulante,* produced by Clara Ibarra and Alexandra Hall, podcast audio, 26:54, https://radioambulante.org/en/audio -en/thephotographer.

Alarcón, Daniel. 2018. "Tolerancia Cero," June 22, in *Radio Ambulante,* produced by Camila Segura, Silvia Viñas, Luis Fernando Vargas, and Daniel Alarcón, podcast audio, 24:28, https://radioambulante.org/en/audio-en/zero-tolerance.

Alarcón, Daniel. 2019. "Una Cadena Humana." November 5, in *Radio Ambulante,* produced by Andrea López-Cruzado, podcast audio, 29:54, https://radioambulante.org /transcripcion/una-cadena-humana-transcripcion.

Alarcón, Daniel. 2020. "Una Ciudad en Dos," February 18, in *Radio Ambulante,* produced by Camila Segura and Luis Fernando Vargas, podcast audio, 30:21, https://radioambulante .org/audio/una-ciudad-en-dos.

Alma DDB. 2011. "A Brave New World of Consumadores: Introducing Young Fusionistas." http://almaad.com/wp-content/uploads/Alma_YellowPaper_Fusionistas1.pdf.

Alvarado, Nicolas. 2013. "NPR's 'Alt.Latino' Celebrates Culture through New Music." *USA Today,* September 15. https://www.usatoday.com/story/life/music/2013/09/15/npr-alt -latino-celebrates-culture-through-new-music/2811823/.

Amaya, Hector. 2013. *Citizenship Excess: Latino/as, Media, and the Nation.* New York: NYU Press.

American Public Media. n.d. "Public Insight Network." APM. https://www.americanpublic media.org/public-insight-network/.

Andrisani, Vincent. 2015. "The Sweet Sounds of Havana: Space, Listening, and the Making of Sonic Citizenship." *Sounding Out* (blog), September 17. https://soundstudiesblog .com/2015/09/17/the-sweet-sounds-of-havana-space-listening-and-the-making-of -sonic-citizenship/.

Ang, Ien. 1991. *Desperately Seeking the Audience.* New York: Routledge.

Anguiano, Jose. 2018. "Soundscapes of Labor and Belonging: Mexican Custodians' Radio Listening Practices at a Southern California University." *Journal of Popular Music Studies* 30 (1–2): 127–54.

Appadurai, Arjun. 1996. *Modernity at Large: Cultural Dimensions of Globalization*. Minneapolis: University of Minnesota Press.

Aspen Institute. 2015. "Reporting on Race in the 21st Century." Annie E. Casey Foundation. August 1. https://www.aecf.org/resources/reporting-on-race-in-the-21st-century/.

Audience Research Analysis (ARA). 1999. "AUDIENCE 98: Public Service, Public Support." https://arapublic.com/ara-research-library.

Bailey, George. 2004. "Free Riders, Givers, and Heavy Users: Predicting Listener Support for Public Radio." *Journal of Broadcasting and Electronic Media* 48 (4): 607–19.

Banet-Weiser, Sarah. 2007. *Kids Rule!: Nickelodeon and Consumer-Citizenship*. Durham, N.C.: Duke University Press.

Baum, Dan. 2006. "Arriba! A Latino Radio Scold Gets the Vote Out." *New Yorker*, October 16. https://www.newyorker.com/magazine/2006/10/23/arriba.

Bauman, Zygmunt. 1996. "From Pilgrim to Tourist: A Short History of Identity." In *Questions of Cultural Identity*, edited by Stuart Hall and Paul Du Gay, 18–36. Thousand Oaks, C.A.: Sage Publications.

Bell, Allan. 1983. "Broadcast News as a Language Standard." *International Journal of the Sociology of Language* 1983 (40): 29–42.

Beltrán, Cristina. 2010. *The Trouble with Unity: Latino Politics and the Creation of Identity*. Oxford: Oxford University Press.

Benson, Rodney and Erik Neveu. 2005. *Bourdieu and the Journalistic Field*. Cambridge: Polity Press.

Berkman, Dave. 1980. "Minorities in Public Broadcasting." Journal of Communication 30 (3): 179–88.

Bernstein, Nina. 2006. "Groundswell of Protests Back Illegal Immigrants." *New York Times*, March 27. https://www.nytimes.com/2006/03/27/us/groundswell-of-protests-back-illegal-immigrants.html.

Bishop, Marlon and Fernanda Echávarri. 2016. "The Strange Death of José de Jesús: Lost in America's Deportation Bureaucracy." *Marshall Project*, July 22. https://www.the marshallproject.org/2016/07/15/the-strange-death-of-jose-de-jesus.

Block, Melissa. 2006. "Spanish D.J. Organizes Immigration-Reform Protests." *National Public Radio*, March 28. https://www.npr.org/templates/story/story.php?storyId=5307593.

Boorstin, Daniel, J. 1971. *The Image: A Guide to Pseudo Events in America*. New York: Atheneum.

Bourdieu, Pierre. 1977. *Outline of a Theory of Practice*. Cambridge: Cambridge University Press.

Bourdieu, Pierre. 1991. *Language and symbolic power*. New York: The New Press.

Bourdieu, Pierre. 1993. *The Field of Cultural Production*. New York: Columbia University Press.

Bourdieu, Pierre. 1990. *On Television*. New York: Columbia University Press.

Brockell, Gillian. 2019. "'Hispanic Invasion': A White Nationalist Version of Texas that Never Existed." *Washington Post*, August 5. https://www.washingtonpost.com/history/2019/08/05/hispanic-invasion-white-nationalist-version-texas-that-never-existed/.

Buckley, Kiera. 2016. "Trump Supporter Says Eugene Rally was 'Electrifying.'" KLCC, May 6. https://www.klcc.org/post/trump-supporter-says-eugene-rally-was-electrifying.

Burnett, John. 2018. "Government Misses Migrant Family Reunification Deadline." *National Public Radio*, July 10. https://www.npr.org/2018/07/10/627821359/government-misses -migrant-family-reunification-deadline.

Carnegie Commission on Educational Television. 1967. *Public Television: A Program for Action*. New York: Bantam Books.

Carnegie Commission on the Future of Public Broadcasting. 1979. *A Public Trust: The Landmark Report of the Carnegie Commission on the Future of Public Broadcasting*. New York: Bantam Books.

Carvin, Andy. 2011. "What is the Origin of the Distinctive NPR Speech Pattern?" Quora, January 4. https://www.quora.com/What-is-the-origin-of-the-distinctive-NPR-speech-pattern.

Casillas, Delores Inés. 2014. *Sounds of Belonging: U.S. Spanish-Language Radio and Public Advocacy*. New York: New York University Press.

Cassidy, Suzanne. 1992. "How the BBC Learns to Say it Right." *New York Times*, August 27. http://www.nytimes.com/1992/08/27/arts/how-the-bbc-learns-to-say-it-right.html.

Castañeda, Mari. 2008. "Re-thinking the U.S. Spanish Language Media Market in an Era of Deregulation." In *Global Communications: Toward a Transnational Political Economy*, edited by Paula Chakravartty and Yuezhi Zhao, 201–18. Lanham, M.D.: Rowman and Littlefield.

Castañeda, Mari. 2016. "Altering the U.S. Soundscape through Latina/o Community Radio." In *Routledge Companion to Latina/o Media Studies*, edited by María Elena Cepeda and Delores Ines Casillas, 110–22. New York: Routledge.

Champagne, Patrick. 2005. "The 'Double Dependency': The Journalistic Field Between Politics and Markets." In *Bourdieu in the Journalistic Field*, edited by Rodney Benson and Erik Neveu, 48–63. Cambridge: Polity Press.

Chávez, Christopher. 2014. "Constructing Latino Consumer-Citizens: An Analysis of Print Advertising in *El Clamor Público* (1855) and *La Opinión* (1926)." *Howard Journal of Communication* 25 (2): 192–210.

Chávez, Christopher. 2015. *Reinventing the Latino Television Viewer: Language, Ideology and Practice*. Lanham, M.D.: Lexington Books.

Chávez, Leo. 2008. *The Latino Threat: Constructing Immigrants, Citizens and the Nation*. Stanford: Stanford University Press.

Contreras, Felix and Jasmine Garsd. 2011. "Mala Rodriguez Shares Her Music and Influences," August 10, in *Alt.Latino*, produced by Anne Hoffman, podcast audio, 30:22, https://www.npr.org/sections/altlatino/2011/07/28/138791341/this-week-on-alt-latino-spanish-rapper-mala-rodriguez-on-her-music-and-influence.

Contreras, Felix and Jasmine Garsd. 2013. "5 Songs About Immigration, in Song Form," April 11, in *Alt.Latino*, produced by Felix Contreras, podcast audio, 25:34, https://www.npr.org/sections/altlatino/2013/04/11/176910510/5-stories-about-immigration-in-song-form.

Contreras, Felix. 2018. "Protesting Trump's Immigration Policy Through Song," June 1, in *Alt.Latino*, produced by Felix Contreras, podcast audio, 45:19, https://www.npr.org/sections/altlatino/2018/07/11/626537505/protesting-trumps-immigration-policy-through-song.

Contreras, Felix. 2017. "Alt.Latino Visits NPR's Newest Podcast: Radio Ambulante," January 25, in *Alt.Latino*, produced by Felix Contreras, podcast audio, 26:50, https://www.npr.org/sections/altlatino/2017/01/25/511396559/alt-latino-visits-nprs-newest-podcast-radio-ambulante.

Corona, Ignacio. 2017. "The Cultural Location/s of (U.S.) Latin Rock." In *The Routledge Companion to Latina/o Media*, edited by María Elena Cepeda and Delores Inés Casillas, 241–58. New York: Routledge.

Corporation for Public Broadcasting. 1978. *A Formula for Change: The Report of the Task Force on Minorities in Public Broadcasting*. Report prepared for the U.S. Department of Health, Education & Welfare National Institute of Education.

Correal, Annie. 2011. "Radio Ambulante." *Transom* (blog), September 6. https://transom.org/2011/radio-ambulante/.

Cruz, Frank and Rita Joiner Soza. 2019. *Straight Out of Barrio Hollywood: The Adventures of Telemundo Co-Founder Frank Cruz, History Professor, TV Anchorman, Network Executive and Public Broadcasting Leader*. Denver: Outskirts Press.

Current. 2014. "David Candow, 'Host Whisperer' and Public Radio Trainer, Dies at 74." *Current*. September 28. https://current.org/2014/09/david-candow-host-whisperer-and-public-radio-trainer-dies-at-74/.

Cwynar, Christopher. 2017. "NPR Music: Remediation, Curation, and National Public Radio in the Digital Convergence Era." *Media, Culture, & Society* 39 (5): 680–96.

Darras, Eric. 2005. "Media Concentration of the Political Order." In *Bourdieu in the Journalistic Field*, edited by Rodney Benson and Erik Neveu, 156–73. Cambridge: Polity Press.

Dávila, Arlene. 2008. *Latino Spin: Public Image and the Whitewashing of Race*. New York: New York University Press.

Dávila, Arlene. 2001. *Latinos, Inc.: The Marketing and Making of a People*. Berkeley: University of California Press.

Davidson, Adam. 2006. "Q&A: Illegal Immigrants and the Economy." *National Public Radio*, March 30, 2006. https://www.npr.org/templates/story/story.php?storyId=5312900.

de Assis, Carolina. 2019. "With 'Listening Clubs,' *Radio Ambulante* Wants to Bring Latin America Closer to the Podcast." *Knight Center for Journalism in the Americas* (blog), April 24. https://knightcenter.utexas.edu/blog/00-20820-listening-clubs-radio-ambulante-wants-bring-latin-america-closer-to-podcast.

De La Cruz, Sonia. 2017. "Latino Airwaves: Radio Bilingüe and Spanish-Language Public Radio." *Journal of Radio and Audio Media* 24 (2): 226–37.

del Valle, José. 2006. "U.S. Latinos, *La Hispanofonía*, and the Language Ideologies of High Modernity." In *Globalization and Language in the Spanish-Speaking World*, edited by Clare Mar-Molinero and Miranda Stewart, 27–46. London: Palgrave McMillan.

Duranti, Alessandro. 1997. *Linguistic Anthropology*. Cambridge: Cambridge University Press.

Dvorkin, Jeffrey. 2005a. "Pronunciamentos: Saying It Right." *National Public Radio*, November 8. http://www.npr.org/templates/story/story.php?storyId=4994862.

Dvorkin, Jeffrey. 2005b. "Why Doesn't NPR Sound More Like the Rest of America?" *National Public Radio*, May 18. http://www.npr.org/templates/story/story.php?storyId=4656584H.

Eco, Umberto. 1997. *Travels in Hyperreality*, trans. William Weaver. New York: Picador.

Edison Research. 2020. "Latino Podcast Listener Report." http://www.edisonresearch.com/wp-content/uploads/2020/06/US-Latino-Podcast-Listener-Report-Edison-Research-English.pdf.

Elliott, Debbie and Carrie Kahn. 2006. "L.A. March Protests Looming Immigration Law." *National Public Radio*. March 25. https://www.npr.org/templates/story/story.php?storyId=5301795.

Engelman, Ralph. 1996. *Public Radio and Television in America: A Political History*. Thousand Oaks, C.A.: Sage Publications.

Falk, Tyler. 2015a. "Drop in Younger Listeners Makes Dent in NPR News Audience." *Current*. October 16. https://current.org/2015/10/drop-in-younger-listeners-makes-dent-in-npr-news-audience/?wallit_nosession=1.

Falk, Tyler. 2015b. "How KPCC worked to grow its Latino audience in Los Angeles." *Current*, July 6. https://current.org/2015/07/how-kpcc-worked-to-grow-its-latino-audience-in-los-angeles/?wallit_nosession=1.

Falk, Tyler. 2015c. "Behind NPR's Decision to Remove Its Branding for a *Latino USA* Episode." *Current*, April 24. https://current.org/2015/04/behind-nprs-decision-to-remove-its-branding-from-a-latino-usa-episode/?wallit_nosession=1.

Falk, Tyler. 2015d. "NPR Aims to Raise Profile of Its Podcasts with Its New Directory." *Current*, April 24. https://current.org/2015/04/behind-nprs-decision-to-remove-its-branding-from-a-latino-usa-episode/?wallit_nosession=1.

Farhi, Paul. 2008. "When this Guy Talks, NPR Listens." *Washington Post*, August 31. http://www.washingtonpost.com/wp-dyn/content/article/2008/08/29/AR2008082900683.html.

Farhi, Paul. 2010. "NPR Has Become a Major Player on the Indie Rock Scene." *Washington Post*, June 6. https://www.washingtonpost.com/wp-dyn/content/article/2010/06/04/AR2010060402177.html.

Farhi, Paul. 2015. "NPR is Graying, and Public Radio is Worried About It." *Washington Post*, November 22. https://www.washingtonpost.com/lifestyle/style/npr-is-graying -and-public-radio-is-worried-about-it/2015/11/22/0615447e-8e48-11e5-baf4-bdf3735 5da0c_story.html?noredirect=on&utm_term=.74a78fe11b1d.

Feld, Steven. 2000. "A Sweet Lullaby for World Music." *Public Culture* 12 (1):145–71.

Félix, Adrián, Carmen González, and Ricardo Ramirez. 2008. "Political Protest, Ethnic Media, and Latino Naturalization." *American Behavioral Scientist* 52 (4): 618–34.

Félix, Maritza. 2017. "Celda 603: La Agonía de un Inmigrante." *Telemundo Arizona*, January 19. https://www.telemundoarizona.com/noticias/local/reportaje-especial-jose-de -jesus-deniz-sahagun/24181/.

Ferguson, Andrew. 2004. "Radio Silence: How NPR Purged Classical Music from Its Airwaves." *Weekly Standard*, June 14.

Fernández-Sande, Manuel. 2014. "Radio Ambulante: Narrative Radio Journalism in the Age of Crowdfunding." In *Radio Audiences and Participation in the Age of Network Society*, edited by Tiziano Bonini and Belén Monclús, 176–94. New York: Routledge.

File, Tom. 2018. "Characteristics of Voters in the Presidential Election of 2016." https:// www.census.gov/content/dam/Census/library/publications/2018/demo/P20-582.pdf.

Fisher, Walter. 1989. *Narration as Human Communication: Toward a Philosophy of Reason, Value, and Action*. Columbia: University of South Carolina Press.

Flaccus, Gillian. 2006. "Spanish-Language Media Credited on Pro-Immigrant Rallies." *Boston Globe*, March 29. http://archive.boston.com/news/nation/articles/2006/03/29 /spanish_language_media_credited_on_pro_immigrant_rallies/.

Flores, Antonio. 2017. "2015, Hispanic Population in the United States Statistical Portrait." *Pew Research Center*. September 18. https://www.pewresearch.org/hispanic/2017/09 /18/2015-statistical-information-on-hispanics-in-united-states/.

Flynn, Meagan. 2020. "Wisconsin Chief Justice Sparks Backlash by Saying Covid-19 Outbreak Is Among Meatpacking Workers, Not the 'Regular Folks.'" *Washington Post*, May 7. https://www.washingtonpost.com/nation/2020/05/07/meatpacking-workers -wisconsin-coronavirus/.

Foucault, Michel. 1984. "The Order of Discourse." In *Language and Politics*, edited by Michael J. Shapiro, 108–38. Oxford: Oxford University Press.

Fraser, Nancy. 1990. "Rethinking the Public Sphere: A Contribution to the Critique of Actually Existing Democracy." *Social Text*, 1990 (25/26): 56–80.

Freedman, Samuel. 2001. "Television/Radio; Public Radio's Private Guru." *New York Times*, November 11. https://www.nytimes.com/2001/11/11/arts/television-radio-public-radio -s-private-guru.html.

Frey, William. 2018. "The U.S. Will Become 'Minority' White in 2045, Census Projects." Brookings Institute. March 14. https://www.brookings.edu/blog/the-avenue/2018/03 /14/the-us-will-become-minority-white-in-2045-census-projects/.

Friess, Steve. 2017. "Why are #PodcastsSoWhite?" Columbia Journalism Review. March 21. https://www.cjr.org/the_feature/podcasts-diversity.php.

Frith, Simon. 2006. "The Industrialization of Popular Music." In *The Popular Music Studies Reader*, edited by Andy Bennet, Barry Shank, and Jason Toynbee, 231–38. London: Routledge.

Futuro Media Group. 2016. "The Futuro Matters: 2016 Annual Report." https://futuromedia group.org/wp-content/uploads/2017/06/The-Futuro-Media-Group-2016-Annual -Report-.pdf.

Futuro Media Group. 2017. "The Futuro Matters: 2017 Annual Report." https://www.futuro mediagroup.org/futuro-media-annual-report/.

Futuro Media Group. 2020. "Latino USA." https://www.futuromediagroup.org/media-property /latino-usa/.

Garbes, Laura. 2017. "How a CPB Task Force Advanced a Prescient Vision for Diversity in Public Radio." *Current*, November 13. https://current.org/author/laura-garbes/?wallit_nosession=1.

Garcia, Matt. 2010. "Social Movements: The Rise of Colorblind Conservatism, and What Comes Naturally." *Frontiers: A Journal of Women's Studies* 31 (3): 49–56.

Garcia-Canclini, Nestor. 2001. *Consumers and Citizens: Globalization and Multicultural Conflicts*. Minneapolis: University of Minnesota Press.

Garcia-Navarro, Lourdes. 2019. "The Media Erased Latinos from the Story." *Atlantic*, August 9. https://www.theatlantic.com/ideas/archive/2019/08/we-must-recognize-hispanics-were-targeted/595783/.

Garcia-Navarro, Lourdes (@lourdesgnavarro). 2019. "This thread is about why I use correct pronunciation of Spanish on-air." Twitter, April 5, 8:20 a.m. https://twitter.com/lourdesgnavarro/status/1114186623694934017.

Garsd, Jasmine and Felix Contreras. 2010. "What Is Latin Alternative Music? And Who Are We?" NPR. June 6. https://www.npr.org/sections/altlatino/2010/08/03/128963147/who-we-are.

Garsd, Jasmine. 2012. "Guest DJ: Calle 13," *Alt.Latino*. Podcast audio, November 21. https://www.npr.org/2010/11/19/131444760/special-guests-calle-13.

Garsd, Jasmine. 2015. "*Happy Cinco de Morrissey!*" June 5, in *Alt.Latino*, produced by Jasmine Garsd, podcast audio, 41:54, https://www.npr.org/sections/altlatino/2015/05/05/404424532/happy-cinco-de-morrissey.

Garsd, Jasmine. 2015. "*Hear 6 Latin American Artists who Rock Indigenous Languages*," October 15, in *Alt.Latino*, produced by Jasmine Garsd, podcast audio, 37:12, https://www.npr.org/sections/altlatino/2015/03/05/390934624/hear-6-latin-american-artists-who-rock-in-indigenous-languages.

Garsd, Jasmine. 2016. "Las Mostras: Fierce Women of Latin Music," June 22, in *Alt.Latino*, produced by Jasmine Garsd, podcast audio, 37:51, https://www.npr.org/sections/altlatino/2016/07/22/486895476/las-mostras-fierce-women-of-latin-music.

Gildea, Terry and Alicia Zuckerman. 2020. "María Hinojosa Will Receive PMJA's 2020 Leo C. Lee Award." *Public Radio Journalists of America*, February 13. https://www.pmja.org/post/maria-hinojosa-will-receive-pmja-s-2020-leo-c-lee-award.

Glass, Ira. 2020. "We Just Won the First Ever Pulitzer Prize for Audio Journalism!" This American Life. May 4. https://www.thisamericanlife.org/about/announcements/we-just-won-the-first-ever-pulitzer-prize-for-audio-journalism.

Golding, Peter and Graham Murdock. 1991. "Culture, Communication, and Political Economy." In *Mass Media and Society*, edited by James Curran and Michael Gurevitch, 15–32. London: Edward Arnold.

Goldsmith, Jill. 2016. "For StoryCorps, Seeking More Diverse Participation Laid Foundation for Growth." *Current*, July 21. https://current.org/2016/07/for-storycorps-seeking-more-diverse-participation-laid-foundation-for-growth/?wallit_nosession=1.

Goodman, David. 2011. *Radio's Civic Ambition: American Broadcasting in the 1930s*. Oxford: Oxford University Press.

Gordon, Ed. 2006. "Rep. Maxine Waters on Blacks and Immigration." NPR. March 30, in *News & Notes*, 7:04, https://www.npr.org/templates/story/story.php?storyId=5311826.

Green, Emily. 2010. "Radio Waves Connect Mexicans in Remote Tlaxiaco with Families in the US," February 22, in *Public Radio International*, 5:19, https://www.pri.org/stories/2019-02-22/radio-waves-connect-mexicans-remote-tlaxiaco-family-us.

Grieco, Elizabeth. 2018. "Newsroom Employees Are Less Diverse than U.S. Workers Overall." Pew Research Center. November 2. https://www.pewresearch.org/fact-tank/2018/11/02/newsroom-employees-are-less-diverse-than-u-s-workers-overall/.

Guerrero, Carolina. 2018. "Radio Ambulante: Breaking the Language Barrier One Story at a Time." Lecture presented at the Drescher Center for the Humanities, Baltimore, M.D., October. https://www.youtube.com/watch?v=5AaaKXnuKNA.

Guerrero, Carolina. 2020. "Radio Ambulante: A Wealth of Latin American Stories." *The UNESCO Courier*. 1: 10–11.

Gurba, Myriam. 2019. "Pendeja, You Ain't Steinbeck: My Bronca with Fake-Ass Social Justice Literature." *Tropics of Meta*, December 12. https://tropicsofmeta.com/2019/12/12/pendeja-you-aint-steinbeck-my-bronca-with-fake-ass-social-justice-literature/.

Gutiérrez, Felix and Jorge Schement. 1979. *Spanish-Language Radio in the Southwestern United States*. Austin: University of Texas, Center for Mexican American Studies.

Guzmán López, Adolfo. 2014. "The Simpsons Inspires Adolfo Guzman-Lopez to Ask, 'What's in a Name?'" KPCC, January 20. https://www.scpr.org/programs/offramp/2014/01/20/35613/the-simpsons-inspires-kpcc-s-adolfo-guzman-lopez-t/.

Habermas, Jürgen. 1970. "Towards a Theory of Communicative Competence." *Inquiry: An Interdisciplinary Journal of Philosophy* 13:360–375.

Hackley, Christopher. 2002. "The Panoptic Role of Advertising Agencies in the Production of Consumer Culture." *Consumption, Markets, and Culture* 5(3): 211–29.

Hajek, Danny. 2015. "The Inauspicious Start to Susan Stamberg's Broadcasting Career." NPR, April 18. https://www.npr.org/2015/04/18/400466804/the-inauspicious-start-to-susan-stambergs-broadcasting-career.

Harper, Hilliard. 1988a. "Latinos Protest NPR Move to Axe *Enfoque*." *Los Angeles Times*, March 3. https://www.latimes.com/archives/la-xpm-1988-03-03-ca-175-story.html.

Harper, Hilliard. 1988b. "NPR Criticized for Canceling Latino Newscast." *Los Angeles Times*, February 27. https://www.latimes.com/archives/la-xpm-1988-02-27-ca-11937-story.htm.

Hardy, Jonathan. 2014. *Critical Political Economy of the Media*. New York: Routledge.

Harvey, Eric. 2014. "Station to Station: The Past, Present, and Future of Streaming Music." *Pitchfork*, April 19. https://pitchfork.com/features/cover-story/reader/streaming/.

Havens, Timothy, Amanda Lotz, and Serra Tinic. 2009. "Critical Media Industry Studies: A Research Approach." *Communication, Culture & Critique* 2 (2): 234–53.

Herman W. Land Associates. 1967. *The Hidden Medium: A Status Report on Educational Radio in the United States*. New York: Herman W. Land Associates.

Hernandez, Tim. 2017. *All They Will Call You*. Tucson: University of Arizona Press.

Hinojosa, María. 2001. "Coverage of Latino Life Is an American Story." *Nieman Reports*, June 15. https://niemanreports.org/articles/coverage-of-latino-life-is-an-american-story/.

Hinojosa, María. 2016. "The Strange Death of José de Jesús." July 15, in *Latino USA*, produced by Fernanda Echávarri, Marlon Bishop, and María Hinojosa, podcast audio, 50:51, https://www.latinousa.org/josedejesus/.

Hinojosa, María. 2017. "Immigrant's Suicide Raises Questions About Safety of Detention Centers," February 20, in *National Public Radio*, https://www.npr.org/2017/02/20/516292239/immigrants-suicide-raises-questions-about-safety-of-detention-centers.

Hinojosa, María. 2019a. "A Conversation with Bernie Sanders," September 20, in *Latino USA*, produced by Miguel Macías, podcast audio, 32:51, https://www.latinousa.org/2019/09/20/sanders/.

Hinojosa, María. 2019b. "A Conversation with Presidential Candidate Julián Castro," February 13, in *Latino USA*, produced by Miguel Macías, podcast audio, 25:52, https://www.latinousa.org/2019/02/13/juliancastro/.

Hinojosa, María. 2019c. "A Conversation with Cory Booker," June 14, in *Latino USA*, produced by Miguel Macías, podcast audio, 28:44, https://www.latinousa.org/2019/06/14/corybooker/.

Hinojosa, María. 2019d. "Seeking Asylum. Seeking to Stay Together," June 25, in *Latino USA*, produced by Katie Schlechter, podcast audio, 22:27, https://www.latinousa.org /2019/06/25/seekingasylum/.

Hinojosa, María. 2019e. "A Child Lost in Translation," May 3, in *Latino USA*, podcast audio, 36:59, https://www.latinousa.org/2019/05/31/childlostintranslation/.

Hinojosa, María. 2019f. "Language Matters: Estrella's Story." *Change Agent*, n.d. https:// changeagent2019.comnetwork.org/2019/language-matters-estrellas-story/.

Hinojosa, María. 2020a. "A Conversation with Elizabeth Warren," February 20, in *Latino USA*, produced by Miguel Macías, podcast audio, 33:28, https://www.latinousa.org /2020/02/20/elizabethwarren/.

Hinojosa, María. 2020b. "A Conversation with Pete Buttigieg," July 26, in *Latino USA*, produced by Miguel Macías and Maggie Freleng, podcast audio, 28:36, https://www.latino usa.org/2019/07/26/petebuttigieg/.

Hinojosa, María. 2020c. "Digging into 'American Dirt,'" January 29, in *Latino USA*, produced by Antonia Cereijido, podcast audio, 58:13, https://www.latinousa.org/2020/01 /29/americandirt/.

Hinojosa, María. 2020d. "Latino-Owned and Without a Lifeline, Small Businesses Struggle to Survive," May 8, in *Latino USA*, produced by Janice Llamoca, podcast audio, 27:22, https://www.latinousa.org/2020/05/08/smallbusinesses/.

Hinojosa, María. 2020e. "Immigrants in ICE Detention Face the Threat of Covid-19," April 7, in *Latino USA*, produced by Miguel Macías and Alissa Escarce, podcast audio, 30:11, https://www.latinousa.org/2020/04/07/immigrantsicecovid19/.

Hoffman, Miles. 2005. *The NPR Classical Music Companion*. New York: Houghton Mifflin.

Holmes, Chuck. 2017. "A Message from WBHM GM Chuck Holmes." WBHM. January 24, https://wbhm.org/2017/wbhm-gm-chuck-holmes-012617/.

Hymes, Dell. 1972. "On Communicative Competence." In *Sociolinguistics: Selected Readings*, edited by J.B. Pride and Janet Holmes, 269–93. Harmondsworth: Penguin.

Ingram, Paul. 2015. "More than 200 Strike at Eloy Detention Center." *Tucson Sentinel*, June 13. http://www.tucsonsentinel.com/local/report/061315_ice_strike/more-than-200 -hunger-strike-eloy-immigration-detention-center/.

Irvine, Judith T. and Suzanne Gal. 2000. "Language Ideology and Linguistic Differentiation." In *Regimes of Language: Ideologies, Polities and Identities*, edited by Paul Kroskrity, 35– 83. Santa Fe: School of American Research Press.

Janssen, Mike and Steve Behrens. 2001. "Bill Siemering Reflects on Launching 'ATC,' Lessons from International Radio Work." Current, May 14. https://current.org/2001/05 /bill-siemering-reflects-on-launching-atc-lessons-from-international-radio-work/ ?wallit_nosession=1.

Jensen, Elizabeth. 2013. "Gambit to Go Independent Opens New Doors for Hinojosa." *Current*, October 10. https://current.org/2013/10/gambit-to-go-independent-opens-new -doors-for-hinojosa/?wallit_nosession=1.

Jensen, Elizabeth. 2018. "NPR's Staff Diversity Numbers, 2017." *National Public Radio*, January 28. https://www.npr.org/sections/publiceditor/2018/01/23/570204215/nprs-staff -diversity-numbers-2017.

Jensen, Elizabeth. 2019a. "New On-Air Source Diversity Data for NPR Show Much Work Ahead." *National Public Radio*, December 17. https://www.npr.org/sections/public editor/2019/12/17/787959805/new-on-air-source-diversity-data-for-npr-shows-much -work-ahead.

Jensen, Elizabeth. 2019b. "NPR's Staff Diversity Numbers, 2019." *National Public Radio*, November 19. https://www.npr.org/sections/publiceditor/2019/11/19/779261818/nprs -staff-diversity-numbers-2019.

Jensen, Elizabeth. 2019c. "You Say Bogota, I Say Bogotá—And That's A Beautiful Thing." *National Public Radio*, April 24. https://www.npr.org/sections/publiceditor/2019/04/24/716379290/you-say-bogota-i-say-bogot-and-thats-a-beautiful-thing.

Jensen, Elizabeth. 2019d. "'Racist,' Not 'Racially Charged': NPR's Thinking on Labeling the President's Tweets." *National Public Radio*, July 23. https://www.npr.org/sections/publiceditor/2019/07/23/744412665/racist-not-racially-charged-npr-s-thinking-on-labeling-the-president-s-tweets.

Jiménez, Carlos. 2019. "Antenna Dilemmas: The Rise of an Indigenous-Language, Low-Power Radio Station in Southern California." *Journal of Radio and Audio Media* 26 (2): 247–69.

Johnson, Fern. 2000. *Speaking Culturally: Language Diversity in the United States*. Thousand Oaks, CA: Sage Publications.

Julia, Megan. 2015. "Autopsy Raises Questions About Eloy Detainees Suicide." *Azcentral.com*, June 24. https://www.azcentral.com/story/news/politics/immigration/2015/06/18/autopsy-questions-eloy-immigration-detainee-suicide-jose-de-jesus-deniz-sahagu/28907315/.

Kellner, Douglas. 2009. "Toward a Critical Media/Cultural Studies." In *Media/Cultural Studies: Critical Approaches*, edited by Rhonda Hammer and Douglas Kellner, 5–24. New York: Peter Lang.

Kern, Jonathan. 2008. *Sound Reporting: The NPR Guide to Audio Journalism and Production*. Chicago: The University of Chicago Press.

Krogstad, Jens Manuel and Mark Hugo López. 2014. "Hispanic Nativity Shift: U.S. Births Drive Population Growth as Immigration Stalls." *Pew Research Center*, April 29. https://www.pewresearch.org/hispanic/2014/04/29/hispanic-nativity-shift/.

Krogstad, Jens Manuel, Mark Hugo López, Gustavo López, Jeffrey Passel, and Eileen Patten. 2016. "Looking Forward to 2016: The Changing Latino Electorate." *Pew Research Center*, January 19. https://www.pewresearch.org/hispanic/2016/01/19/looking-forward-to-2016-the-changing-latino-electorate/.

Kumanyika, Chenjerai. 2015. "Vocal Color in Public Radio." *Transom* (blog), January 22. http://transom.org/2015/chenjerai-kumanyika/.

Kun, Josh. 2005. *Audiotopia: Music, Race, and America*. Berkeley and Los Angeles: University of California Press.

Lapin, Andrew. 2013. "Radio Ambulante Partners with PRI to Produce English-Language Content." *Current*, September 26. https://current.org/2013/09/radio-ambulante-partners-with-pri-to-produce-english-language-content/.

Lapin, Andrew. 2016. "Radio Ambulante Joins NPR Distribution as Network's First Spanish Language Podcast." *Current*, November 15. https://current.org/2016/11/radio-ambulante-joins-npr-podcast-distribution-as-networks-first-spanish-language-program/.

Latino Public Radio Consortium. 2015. "How Southern California Public Radio Opened its Doors to Latinos and Became the Most Listened to Public Station in Los Angeles: A Case Study." https://latinopublicradioconsortium.files.wordpress.com/2015/01/brown-paper-catalog-6-final.pdf.

Latino USA. 2020. "Latino USA and PRX Announce New Partnership." *Latino USA*, June 30. https://www.latinousa.org/2020/06/30/latinousaprx/.

Lepore, Jill. 2010. "Untimely: What Was at Stake in the Spat Between Henry Luce and Harold Ross?" *New Yorker*, April 12. https://www.newyorker.com/magazine/2010/04/19/untimely-jill-lepore.

Lepore, Jill. 2020. "The Last Time Democracy Almost Died." *New Yorker*, February 3. https://www.newyorker.com/magazine/2020/02/03/the-last-time-democracy-almost-died.

Leslie, Deborah. 1995. "Global Scan: The Globalization of Advertising Agencies, Concepts, and Campaigns." *Economic Geography* 71 (4): 402–26.

Libbey, Ted. 2006. *The NPR Listener's Encyclopedia of Classical Music*. New York: Workman Publishing Co.

Lippi-Green, Rosina. 2011. *English with an Accent: Language, Ideology and Discrimination in the United States*, 2nd ed. New York: Routledge.

Lipsitz, George. 1994. *Dangerous Crossroads: Popular Music, Postmodernism and the Poetics of Place*. London: Verso.

Livingston, Gretchen. 2010. "Latinos and Digital Technology, 2010." *Pew Research Center*. February 11. http://www.pewhispanic.org/2011/02/09/latinos-and-digital-technology-2010/.

Lloyd, James A. 1926. "Broadcast English." In *The English Language*, Volume 2: *Essays by Linguists and Men of Letters, 1858–1964*, edited by W.F. Bolton and David Crystal, 100–13. Cambridge: Cambridge University Press.

Lochte, Bob. 2003. "U.S. Public Radio–What Is It? And for Whom?" In *More than A Music Box: Radio Cultures and Communities in a Multi-Media World*, edited by Andrew Crisell, 39–56. New York: Berghahn Books.

Lopez, Hugo, Ana Gonzalez-Barrera, and Jens Manuel Krogstad. 2018. "More Latinos Have Serious Concerns About Their Place in America Under Trump." *Pew Research Center*. October 25. https://www.pewresearch.org/hispanic/2018/10/25/more-latinos-have-serious-concerns-about-their-place-in-america-under-trump/.

Lornell, Kip. 2004. *The NPR Curious Listener's Guide to American Folk Music*. New York: The Berkeley Publishing Group.

Los Angeles Times. 1948. "Plane Crash Kills 32 Near Coalinga: Craft Taking Farm Workers to Be Deported." *Los Angeles Times*, January 29.

Los Angeles Times. 2019. "Introducing . . . The Battle of Los Angeles." *Los Angeles Times*, October 29. https://www.latimes.com/california/story/2019-10-15/prop-187-this-is-california-battle-podcast.

Loviglio, Jason. 2008. "Sound Effects: Gender, Voice and the Cultural Work of NPR." *The Radio Journal: International Studies in Broadcast and Audio Media* 5 (2):67–81.

Loviglio, Jason. 2012. "US Public Radio, Social Change, and the Gendered Voice." In *Electrified Voices: Medial, Socio-historical, and Cultural Aspects of Voice Transfer*, edited by Dmitri Zakharine and Nils Meise, 137–46. Göttingen: V&R Unipress.

Loviglio, Jason. 2013. "Public Radio in Crisis." In *Radio's New Wave: Global Sound in the Digital Era*, edited by Jason Loviglio and Michele Hilmes, 24–42. New York: Taylor & Francis.

Lowey, Nita. 2020. "From Rep. Nita Lowy: Trump vs. NPR." *New York Times*, February 19. https://www.nytimes.com/2020/02/19/opinion/letters/national-public-radio.html.

Luiselli, Valeria. 2019. "The Wild West Meets the Southern Border." *New Yorker*, June 10. https://www.newyorker.com/magazine/2019/06/10/the-wild-west-meets-the-southern-border.

Macadam, Alison. 2015. "How a Long Audio Story Differs from a Short One." *NPR Training*. March 9. https://training.npr.org/2015/03/09/how-a-long-audio-story-is-different-from-a-short-one/.

MacArthur Foundation. 2020. "Grant Search: Radio Ambulante." MacArthur Foundation, accessed June 1, 2020. https://www.macfound.org/grantees/10416/.

Mahony, Roger. 2006. "Called by God to Help." *New York Times*, March 22. https://www.nytimes.com/2006/03/22/opinion/called-by-god-to-help.html.

Maloy, Simon. 2006. "NPR's Williams on Immigration Protesters: 'These Kids Don't Know Anything.'" *Media Matters*, March 30. https://www.mediamatters.org/fox-news/nprs-williams-immigration-protesters-these-kids-dont-know-anything.

Marchetti, Dominique. 2005. "Subfields of Specialized Journalism." in *Bourdieu in the Journalistic Field*, edited by Rodney Benson and Erik Neveu, 64–84. Cambridge: Polity Press.

Marcum, Diana. 2013. "Decades After Crash, Names of 28 Deportees Are Read Aloud." *Los Angeles Times*, September 2. https://www.latimes.com/local/lanow/la-xpm-2013-sep-02-la-me-ln-deportees-name-read-aloud-20130902-story.html.

Margolick, Davide. 2012. "National Public Rodeo." *Vanity Fair,* January 18. https://www
.vanityfair.com/news/business/2012/01/National-Public-Rodeo.

Martin, María. 2017. "Crossing Borders." In *Reality Radio: Telling True Stories in Sound,*
2nd ed., edited by John Biewen and Alexa Dilworth, 204–11. Chapel Hill: University of
North Carolina Press.

Martin, María. 2020. *Crossing Borders, Building Bridgers: A Journalist's Heart in Latin
America.* San Antonio: Conocimientos Press, LLC.

Martínez Wood, Jamie. 2007. *Latino Writers and Journalists (A to Z of Latino Americans).*
New York: Infobase Publishing.

Mason, Jennifer. 2002. *Qualitative Researching.* Thousand Oaks, C.A.: Sage Publishing.

Massey, Douglas and Nancy Denton. 1993. *American Apartheid: Segregation and the Making of the American Underclass.* Cambridge, MA: Harvard University Press.

McCauley, Michael. 2005. *NPR: The Trials and Triumphs of National Public Radio.* New
York: Columbia University Press.

McChesney, Robert. 2008. *The Political Economy of Media: Enduring Issues, Emerging Dilemmas.* New York: Monthly Review Press.

McCourt, Tom. 1999. *Conflicting Communication Interests in America: The Case of National Public Radio.* Westport, C.T.: Praeger Publishers.

McCracken, Grant. 1988. *The Long Interview.* Newberry Park, C.A.: Sage Publications.

McDowell, John, María Herrera-Sobek, and Rodolfo Cortina. 1994. "Hispanic Oral Tradition: Form and Content." In *Handbook of Hispanic Cultures in the United States,* edited
by Francisco Lomelí, 218–25. Houston: Arte Publico Press.

McKee, Alan. 2003. *Textual Analysis: A Beginner Guide.* Thousand Oaks, C.A.: Sage
Publications.

Memmott, Mark. 2014. "Wondering How to Say That Name? Remember, Help Is Near."
National Public Radio, October 21. https://www.npr.org/sections/memmos/2014/10
/21/605359921/wondering-how-to-say-that-name-remember-help-is-near.

Memmot, Mark. 2017. "Update: Guidance on Immigration." *National Public Radio,* July
25. https://www.npr.org/sections/memmos/2017/07/25/606412711/update-guidance
-on-immigration.

Mignolo, Walter. 2005. *The Idea of Latin America.* Malden, M.A.: Blackwell Publishing.

Milroy, James. 2001. "Language Ideologies and the Consequences of Standardization." *Journal of Sociolinguistics* 5 (4): 530–55.

Milroy, Lesley. 2001. "Britain and the United States: Two Nations Divided by the Same
language (and Different Language Ideologies)." *Journal of Linguistic Anthropology* 10
(1): 56–89.

Milroy, Lesley and James Milroy. 2012. *Authority in Language: Investigating Standard English.* New York: Routledge.

Mora, Cristina. 2014. *Making Hispanics: How Activists, Bureaucrats, and Media Constructed
a New American.* Chicago: University of Chicago Press.

Moreno, Marisel and Thomas Anderson. 2014. "'I am an American Writer': An Interview
with Daniel Alarcón." *MELUS* 39 (4): 186–206.

Muggleston, Lynda. 2008. "Spoken English and the BBC: In the Beginning." *Arbeiten aus
Anglistik Und Amerikanist* 33 (2): 197–216.

Mullin, Benjamin. 2015. "Inside NPR's Podcasting Strategy." *Poynter,* March 30. https://
www.poynter.org/reporting-editing/2015/inside-nprs-podcasting-strategy/.

National Public Media. 2017. "Latino USA: Sharing Latino Voices and Perspectives Across
Platforms." National Public Media.

National Public Media. 2020. "NPR Audience Profile." National Public Media, n.d. https://
www.nationalpublicmedia.com/audience/.

National Public Radio. n.d. "Our Mission and Vision." *National Public Radio,* n.d. https://
www.npr.org/about-npr/178659563/our-mission-and-vision.

National Public Radio. n.d. "NPR Training: Storytelling Tips and Best Practices." *National Public Radio*, n.d. https://training.npr.org.

National Public Radio. 2012. "Ana Tijoux: Addressing Global Unrest in Rhyme." *National Public Radio*, February 12. https://www.npr.org/transcripts/146694189.

National Public Radio. 2017. "Debut Broadcast of *All Things Considered* (May 3, 1971)." *National Public Radio*, March 29. https://one.npr.org/?sharedMediaId=521838960:521 841616.

National Public Radio. 2018. "NPR Maintains Highest Ratings Ever." *National Public Radio*, March 28. https://www.npr.org/about-npr/597590072/npr-maintains-highest-ratings -ever.

National Public Radio. 2019. "NPR Board Welcomes Newly Elected Members." *National Public Radio*, September 16. https://www.npr.org/about-npr/760503976/npr-board -welcomes-newly-elected-directors.

National Public Radio. 2020a. "NPR Stations and Public Media." *National Public Radio*, n.d. https://www.npr.org/about-npr/178640915/npr-stations-and-public-media.

National Public Radio. 2020b. "Public Radio Finances." *National Public Radio*, n.d. https:// www.npr.org/about-npr/178660742/public-radio-finances.

National Public Radio. 2020c. "Podcast Directory: *Radio Ambulante*." *National Public Radio*, n.d. https://www.npr.org/podcasts/510315/radio-ambulante.

NBC News. 2016. "Humanizing America: Young and Latino." *NBC News*, March 16. https:// www.nbcnews.com/video/humanizing-america-young-latino-644906563978.

Newman, Michael. 2009. "Indie Culture: In Pursuit of the Authentic Autonomous Alternative." *Cinema Journal*, 48 (3): 16–34.

New York Times. 1948. "32 Killed in Crash of Charter Plane: California Victims Include 28 Mexican Workers Who Were Being Deported." *New York Times*, January 29.

New York Times. 1988. "U.S. Hispanic Population is Up 34% Since 1980." *The New York Times*, September 7.

Nickson, Chris. 2004. *The Curious Listener's Guide to World Music*. New York: Berkley Publishing Group.

Nielsen. 2018. "Audio Today 2018: A Focus on Black and Hispanic Audiences." Nielsen, July 11. https://www.nielsen.com/us/en/insights/report/2018/audio-today-a-focus-on -black-and-hispanic-audiences/.

Nieman Reports. 2001. "Coverage of Latino Life Is an American Story." *Nieman Reports*, June 15. https://niemanreports.org/articles/coverage-of-latino-life-is-an-american-story/.

Nieman Reports. 2020. "Learn How Radio Ambulante, a Spanish Language Podcast, Built a Devoted Audience." *Nieman Reports*, March 16. https://niemanreports.org/articles /wradio-ambulante-spanish-language-podcast-finds-a-devoted-audience/.

Noel, Hannah. 2017. "Imagining Postville: National Public Radio and the Discourse of Latina/o Representation." In *Routledge Companion to Latina/o Media*, edited by María Elena Cepeda and Delores Ines Casillas, 209–22. New York: Routledge.

Olivas, Daniel. 2020. "Yes, Latinx Writers Are Angry About American Dirt—And We Will Not Be Silent." *Guardian*, January 30. https://www.theguardian.com/commentisfree /2020/jan/30/american-dirt-book-controversy-latinx-writers-angry.

Oregon Humanities Center. 2017. "From the Frontlines: A Conversation with María Hinojosa." Lecture presented at University of Oregon, Eugene, O.R., October 9. http://media .uoregon.edu/channel/archives/12228.

Ortiz, Renato. 1994. *Mundialização e Cultura*. São Paulo: Brasiliense.

Paredes, Américo. 1958. *With His Pistol in His Hand: A Border Ballad and Its Hero*. Austin: University of Texas Press.

Patten, Eileen. 2016. "The Nation's Latino Population Is Defined by Its Youth." *Pew Research Center*, April 20. https://www.pewresearch.org/hispanic/2016/04/20/the-nations-latino -population-is-defined-by-its-youth/.

Pineda, Dorany. 2020. "As the 'American Dirt' Backlash Ramps Up, Sandra Cisneros Doubles-Down on her Support." *Los Angeles Times*, January 29. https://www.latimes .com/entertainment-arts/books/story/2020-01-29/sandra-cisneros-breaks-silence -american-dirt.

Powell, Tracie. 2015. "Are Podcasts the New Path to Diversifying Public Radio?" *Columbia Journalism Review*, May 22. https://www.cjr.org/analysis/are_podcasts_the_new_path _to_diversifying_public_radio.php.

Pretsky, Holly. 2017. "NPR in Spanish: Approaching Content for a Bilingual Audience." *National Public Radio*, December 14. https://www.npr.org/sections/publiceditor/2017/12 /14/570848670/npr-in-spanish-approaching-content-for-a-bilingual-audience.

PRX. 2020. "About PRX." PRX. n.d. https://www.prx.org/company/about/.

Quah, Nicholas. 2020. "El Hilo: A New Spanish Language News Podcast." *Hot Pod* (blog), March 31. https://hotpodnews.com/el-hilo-a-new-spanish-language-news-podcast/.

Radio Ambulante. 2012. "Welcome to Radio Ambulante." Video, 2:47. n.d., https://vimeo .com/35644222.

Radio Ambulante. 2014. "Las Voces of Radio Ambulante." Video, 1:43. n.d., https://vimeo .com/91046520.

Regatao, Gisele. 2018. "The Many Voices of Journalism." *Columbia Journalism Review*, Fall. https://www.cjr.org/special_report/journalism-accents-radio.php.

Regatao, Gisele. 2019. "American Icons: Crossroad Blues." Public Radio International, February 28. https://www.pri.org/stories/2019-02-28/american-icons-cross-road-blues.

Resnikoff, Paul. 2016. "Two-thirds of All Music Comes from Just 3 Companies." *Digital Music News*, August 3. https://www.digitalmusicnews.com/2016/08/03/two-thirds -music-sales-come-three-major-labels/.

Roberts, Cokie. 2010. *This is NPR: The First Forty Years*. San Francisco: Chronicle Books.

Rohter, Larry. 2013. "A Writer Thrives in Two Cultures." *New York Times*, November 13. https://www.nytimes.com/2013/11/14/books/daniel-alarcon-is-a-hot-talent-in-peru -and-america.html.

Rosa, Jonathan. 2016. "Standardization, Racialization, Languagelessness: Raciolinguistic Ideologies Across Communicative Contexts." *Journal of Linguistic Anthropology* 26:162–83.

Rosaldo, Renato. 1993. *Culture & Truth: Remaking Social Analysis*. Boston: Beacon Press.

Samaha, Albert and Katie Baker. 2020. "Smithfield Foods Is Blaming 'Living Circumstances in Certain Cultures' for One of America's Largest COVID-19 Clusters." *Buzzfeed*, April 20. https://www.buzzfeednews.com/article/albertsamaha/smithfield-foods-corona virus-outbreak.

Sánchez, George. 1995. *Becoming Mexican-American: Ethnicity, Culture, and Identity in Chicano Los Angeles, 1900–1945*. Oxford: Oxford University Press.

Sanders, Sam. 2020. "Reckoning with Race in Journalism," July 14, 2020, in *It's Been a Minute with Sam Sanders*, produced by Andrea Gutiérrez and Anjuli Sastry, podcast audio, 42:43, https://www.npr.org/2020/07/10/889773113/reckoning-with-race-in -journalism.

Schmidt Camacho, Alicia. 2008. *Migrant Imaginaries: Latino Cultural Politics in the U.S.- Mexico Borderlands*. New York: NYU Press.

Schoenberg, Loren. 2002. *The NPR Curious Listener's Guide to Jazz*. New York: Berkley Publishing Group.

Schultz, Ida. 2007. "The Journalistic Gut Feeling: Journalistic Doxa, News Habitus, and Orthodox News Values." *Journalism Practice* 2:190–207.

Schumacher-Matos, Edward. 2012. "Six National Leaders and Experts Look at Diversity at NPR." *National Public Radio*, April 30. https://www.npr.org/sections/publiceditor/2012 /04/30/151304276/six-national-leaders-and-experts-look-at-diversity-at-npr.

Schumacher-Matos, Edward. 2014. "Race at NPR, and the End of Tell Me More." *National Public Radio*, June 30. https://www.npr.org/sections/publiceditor/2014/06/30/325193 324/race-at-npr-and-the-end-of-tell-me-more.

Schwyter, Jürg. 2016. *Dictating to the Mob: History of the BBC Advisory Committee on Spoken English*. Oxford: Oxford University Press.

Shank, Barry. 2014. *The Political Force of Musical Beauty*. Durham, N.C.: Duke University Press.

Shapiro, Ari. 2006. "Do Illegal Immigrants Burden the Justice System?" *National Public Radio*, April 27. https://www.npr.org/templates/story/story.php?storyId=5365863.

Siemering, Bill. 1970. "National Public Radio Purposes, 1970." *Current*, May 17. https://current.org/2012/05/national-public-radio-purposes/.

Smith, Timothy and Michael Tilson Thomas. 2002. *The Curious Listener's Guide to Classical Music*. New York: Berkley Publishing Group.

Smith-Shomade, Beretta. 2008. *Pimpin' Ain't Easy: Selling Black Entertainment Television*. New York: Routledge.

Socolovsky, Jerome. 2019. "Pronounce Like a Polygot: Saying Foreign Names On-Air." *National Public Radio*, April 30. https://training.npr.org/2019/04/30/pronounce-like-a-polyglot-saying-foreign-names-on-air/.

Sounding Out. 2014. "Radio Ambulante: A Radio That Listens." *Sounding Out* (blog), May 8. https://soundstudiesblog.com/2014/05/08/radio-ambulante-a-radio-that-listens/.

Spotswood, Richard. 1991. *Ethnic Music on Records: A Discography of Ethnic Recordings Produced in the United States, 1893–1942*. Urbana-Champaign: University of Illinois Press.

Stamberg, Susan. 2012. *NPR American Chronicles: Women's Equality*. Prince Frederick, M.D.: Highbridge.

Stavitsky, Alan. 1995. "'Guys in Suits with Charts': Audience Research in U.S. Public Radio." *Journal of Broadcasting and Electronic Media* 39 (2):177–89.

Sterling, Christopher, Cary O'Dell. 2011. *Biographical Encyclopedia of American Radio*. New York: Routledge.

Straubhaar, Joseph. 1991. "Beyond Media Imperialism: Assymetrical Interdependence and Cultural Proximity." *Critical Studies in Media Communication* 8 (1): 39–59.

StoryCorps. 2020. "StoryCorps Historias." StoryCorps, accessed April 24, 2020, https://story corps.org/discover/historias/.

Stuart, Tessa. 2012a. "How KPCC's quest for Latino listeners doomed the Madeline Brand Show." *LA Weekly*, November 1. https://www.laweekly.com/news/how-kpccs-quest-for-latino-listeners-doomed-the-madeleine-brand-show-2611923.

Stuart, Tessa. 2012b. "A. Martinez hired at KPCC: The Madeline Brand Show Goes National." *LA Weekly*, August 17. https://www.laweekly.com/news/a-martinez-hired-at-kpcc-the-madeleine-brand-show-gears-up-to-go-national-2387049.

Suarez, Ray. 2013. *Latino Americans: The 500 Year Legacy That Shaped a Nation*. New York: Penguin Books, Ltd.

Suarez, Ray. 2014. "Latino Americans: The 500 Year Legacy That Shaped a Nation." Lecture presented at the Library of Congress, Washington, D.C., February. https://www.loc.gov/item/webcast-6274.

Tamasi, Susan and Lamant Antieau. 2015. *Language and Linguistic Diversity in the U.S.: An Introduction*. New York: Routledge.

Tatum, Charles. 2013. "Latino USA." In *Encyclopedia of Latino Culture: From Calaveras to Quinceañeras*, edited by Charles Tatum, 78–786. Santa Barbara, C.A.: ABC-CLIO.

Tavares, Frank. 1989. "Myth and the Minority Audience: Black and Hispanic Use of Public and Commercial Radio." *CPB Research Notes*, No. 27A.

Thomas, Thomas J. and Theresa R. Clifford. 1988. "AUDIENCE 88: Issues and Implications." Audience Research Analysis. January 1.

Thornton, Sarah. 1996. *Club Cultures: Music, Media, and Subcultural Capital*. Middletown, C.T.: Wesleyan University Press.

Tovares, Joseph. 2016. "Drive for Diversity Demands Courage, Commitment." *Current*, July 21. https://current.org/2016/07/drive-for-diversity-demands-courage-commitment/?wallit_nosession=1.

Tovares, Raúl. 2000. "*Latino USA*: Constructing a News and Public Affairs Radio Program." *Journal of Broadcasting and Electronic Media* 44 (3): 471–86.

Tynan, A. Caroline. and Jennifer Drayton. 2020. "Market Segmentation." *Journal of Marketing Management* 2 (3): 301–35.

Urciuoli, Bonnie. 1996. *Exposing Prejudice: Puerto Rican Experience of Language, Race, and Class*. Boulder, C.O.: Westview Press.

Urry, John. 2002. *The Tourist Gaze, 2nd Edition*. Thousand Oaks, C.A.: Sage Publications.

U.S. Census. 1999. "The Hispanic Population in the United States." https://www.census.gov/prod/2000pubs/p20-525.pdf.

U.S. Census. 2012. "Percent Hispanic of the US Population: 1970–2050." https://www.census.gov/newsroom/cspan/hispanic/2012.06.22_cspan_hispanics_5.pdf.

U.S. Census. 2006. "Hispanics in the United States." https://www.census.gov/population/www/socdemo/files/Internet_Hispanic_in_US_2006.pdf.

U.S. Census. 2020. "Quickfacts: United States." Census.gov, accessed January 5, 2019. https://www.census.gov/quickfacts/fact/table/US/RHI725219.

U.S. Congress. 1988. Public Telecommunications Act of 1988: Hearing Before the Subcommittee on Communications of the Committee on Commerce, Science, and Transportation, United States Senate, One Hundredth Congress, Second Session on S. 2114.

Valdivia, Angharad. 2010. *Latinas/os And the Media*. Malden, M.A.: Polity Press.

Walker, Jesse. 2017. "With Friends Like These: Why Community Radio Does Not Need the Corporation for Public Broadcasting." Cato Policy Analysis No. 277.

Walker, Jesse. 2019. "The Podcast Revolution." *Reason*, September. https://reason.com/2019/08/25/the-podcast-revolution/.

Watanabe, Teresa and Hector Becerra. 2006a. "500,000 Pack Streets to Protest Immigration Bills." *Los Angeles Times*, March 26. https://www.latimes.com/archives/la-xpm-2006-mar-26-me-immig26-story.html.

Watanabe, Teresa and Hector Becerra. 2006b. "How DJ's Put 500,000 in Motion." *Los Angeles Times*, March 28. https://www.latimes.com/archives/la-xpm-2006-mar-28-me-march28-story.html.

Wayne, Teresa. 2015. "NPR Voice Has Taken Over the Airwaves." *New York Times*, October 24. https://www.nytimes.com/2015/10/25/fashion/npr-voice-has-taken-over-the-air waves.html.

Western, Tom. 2020. "Listening with Displacement: Sound, Citizenship, and Disruptive Representations of Migration." Migration and Society 3 (1): 294–310.

Wilkinson, Kenton. 2016. *Spanish-Language Television in the United States: Fifty Years of Development*. New York: Routledge.

Women's Leadership Accelerator. 2017. "Carolina Guerrero: Bio" August 25. https://journalists.org/profiles/carolina-guerrero/.

Zepeda-Millán, Chris. 2017. *Latino Mass Mobilization: Immigration, Racialization, and Activism*. Cambridge: Cambridge University Press.

ABOUT THE AUTHOR

CHRISTOPHER CHÁVEZ is an associate professor in the School of Journalism and Communication at the University of Oregon and director of the university's Center for Latina/o and Latin American Studies. He holds PhD in communications from the University of Southern California. His research lies at the intersection of Latinx media, globalization, and culture. He is author of *Reinventing the Latino Television Viewer: Language Ideology and Practice* and is co-editor of *Identity: Beyond Tradition and McWorld Neoliberalism*. Prior to his doctoral research, Professor Chávez worked as an advertising executive at advertising agencies in San Francisco, Los Angeles, and Boston.